耐火材料实用汉英词典丛书

耐火材料制品
实用术语汉英词典

主　编　马成良　陈留刚
副主编　李素平　原会雨　赵　飞　穆元冬

北　京
冶 金 工 业 出 版 社
2024

内 容 提 要

本书共7章，首先介绍了耐火材料制品的定义及按照不同方法对耐火材料制品进行分类，然后按照化学成分及制备方法分类分别介绍了酸性耐火材料制品、碱性耐火材料制品、中性耐火材料制品、特种耐火材料制品和不定形耐火材料制品等涉及的术语词组、术语句子、术语段落的汉英对照，最后介绍了耐火材料制品常用词汇术语查询，包括汉英对照、英汉对照。

本书可作为无机非金属材料（耐火材料）和冶金工程等专业的研究生、本科生，工程技术人员及对外贸易人员学习和查阅用书，也可供相关专业教师参考使用。

图书在版编目(CIP)数据

耐火材料制品实用术语汉英词典/马成良，陈留刚主编．—北京：冶金工业出版社，2024.1

(耐火材料实用汉英词典丛书)

ISBN 978-7-5024-9721-7

Ⅰ.①耐… Ⅱ.①马… ②陈… Ⅲ.①耐火材料—产品—名词术语—词典—汉、英 Ⅳ.①TQ175.7-61

中国国家版本馆CIP数据核字(2024)第017625号

耐火材料制品实用术语汉英词典

出版发行	冶金工业出版社	**电　　话**	(010)64027926
地　　址	北京市东城区嵩祝院北巷39号	**邮　　编**	100009
网　　址	www.mip1953.com	**电子信箱**	service@mip1953.com

责任编辑　于昕蕾　美术编辑　彭子赫　版式设计　郑小利
责任校对　郑　娟　李　娜　责任印制　禹　蕊
北京印刷集团有限责任公司印刷
2024年1月第1版，2024年1月第1次印刷
710mm×1000mm　1/16；13.5印张；271千字；206页
定价81.00元

投稿电话　(010)64027932　投稿信箱　tougao@cnmip.com.cn
营销中心电话　(010)64044283
冶金工业出版社天猫旗舰店　yjgycbs.tmall.com
(本书如有印装质量问题，本社营销中心负责退换)

前　言

耐火材料是保障高温工业安全高效运行不可缺少的重要支撑材料，也是关系到国民经济发展的重要基础材料。耐火材料的应用涉及钢铁、有色金属、建材、机械、化工、环保及航空航天等国民经济重要支柱产业，这些支柱产业的发展与耐火材料工业技术的发展息息相关。

我国是世界耐火材料第一生产和消耗大国，近年耐火材料研究开发和工业应用水平随着我国高温工业的发展和技术进步得到了全面提升。随着人类社会的发展、高温工业技术的进步及低碳经济时代的到来，耐火材料性能、功能及寿命等迎来了一系列的新要求，这极大地推动了我国耐火材料研究方法、工业技术和材料体系的创新。目前，我国耐火材料已实现从跟跑到并跑甚至领跑的跨越，生产出了许多结构功能化、长寿化、绿色化的新产品。

随着我国耐火材料行业的快速发展，与耐火材料相关的国际学术交流和国际贸易等日益增多。目前国内从事耐火材料研发、生产、应用、服务以及国际贸易的机构和人员众多，每年发表上千篇耐火材料方面的学术论文、标准规范、国家/国际专利、展览海报、宣传飞页，以及相关的合同条文等。因此在科研成果撰写和产品销售宣传等各个层面，都需要科研单位和企业界人士精准掌握耐火材料实用术语及对应的英文表述。通过此书编写，旨在让读者掌握耐火材料不同体系原料规范的汉语和英语表述方法，提高读者的英文表述能力，提升耐火材料从业人员的整体水平和英汉互译素质，为耐火材料行业的国际化发展贡献力量。

本书章节安排中设置了术语词组、术语句子和术语段落，方便读者查阅和学习。本书共分为 7 章，由郑州大学材料科学与工程学院、河南省高温功能材料重点实验室马成良教授和陈留刚副教授担任主编，郑州大学材料科学与工程学院李素平副教授、原会雨副教授、赵飞博士和穆元冬博士担任副主编，参编人员还有郑州大学李烨和史幸福，濮耐高温材料（集团）股份有限公司刘国威和闫光辉。编写工作分工如下：马成良、陈留刚、刘国威和闫光辉负责第1、5、7章，赵飞、史幸福负责第2章，原会雨负责第3章，马成良、李素平和陈留刚负责第4章，穆元冬、李烨负责第6章。在本书的编写过程中，得到了国内外多名耐火材料领域专家教授的指导和帮助，在此一并表示感谢。

限于编者水平和时间仓促，以及当前耐火材料行业的快速发展，在内容和编排上可能会有欠妥和未覆盖之处，敬请广大读者批评指正。

马成良　陈留刚

2023年8月于郑州

目　　录

1 绪 论

按国际标准，耐火材料定义为：化学与物理性质允许其在高温环境下使用的非金属材料与产品（并不排除含有一定比例的金属）。中国标准沿用ISO标准，规定耐火材料是指物理与化学性质适宜于在高温下使用的非金属材料，但不排除某些产品可含有一定量的金属材料。

耐火材料制品的分类方法很多，按照不同的标准存在着不同的划分方式。耐火材料制品按照化学性质可以分为三大类，分别是酸性耐火材料制品、碱性耐火材料制品和中性耐火材料制品。酸性耐火材料制品的主要成分是二氧化硅。这类耐火材料制品在高温下具有较好的耐酸性能，适用于酸性环境下的耐火结构，如炉内衬、烟道等。碱性耐火材料制品通常以氧化镁、氧化钙或二者共同作为主要成分。碱性耐火材料制品对碱性环境具有较好的稳定性，常用于含有碱性成分的工业过程中，如水泥窑、玻璃窑等。中性耐火材料制品在高温下不与酸性耐火材料、碱性耐火材料、酸性或碱性渣或溶剂发生明显的化学反应。中性耐火材料制品具有较广泛的适用范围，可用于多种工业场合，如铁炉、钢炉等。

按照交货形态，可以将耐火材料制品分为定形制品和不定形制品。定形制品是指具有固定形状的耐火砖和保温砖，它们进一步分为致密定形制品和保温定形制品。致密定形制品是指真气孔率低于45%的制品，而保温定形制品则指真气孔率高于45%的制品。定形耐火制品又可以细分为标准形状的砖块以及异形砖等。不定形耐火材料是指由骨料、细粉、结合剂与添加剂组成的混合物，以交货状态直接使用，或加入一种或多种不影响其耐火度的合适液体后使用的耐火材料，也称为散装耐火材料。它们以散料状的形式送到现场，以浇注、捣打、涂抹、喷射、振动、投射或挤压等方式施工。

按照耐火材料中各组分（颗粒、细粉）之间的结合形式，可将耐火材料分为水化结合、陶瓷结合、化学结合、有机结合和树脂结合等。水化结合是指在常温下，通过某种细粉与水发生化学反应产生凝固和硬化而形成的结合，常见于耐火浇注料中，如氧化镁结合浇注料、水泥结合浇注料等。陶瓷结合是指在一定温度下由于烧结或液相形成而产生的结合，常见于烧成制品中。化学结合是指在室温或更高的温度下通过化学反应（不是水化反应）产生硬化形成的结合，常见于各种不烧制品中。有机结合是指在室温或稍高温度下借助有机物质产生硬化而形成的结合，常见于不烧制品中。树脂结合是指含有树脂的耐火材料在较低温度下加热，由于树脂固化、炭化而产生的结合，常见于含碳耐火材料中。

耐火材料按照化学矿物组成可分为硅石耐火材料、铝硅酸盐耐火材料、镁质耐火材料、镁尖晶石质耐火材料、镁铬质耐火材料、镁白云石质耐火材料、白云石质耐火材料和碳复合耐火材料等。硅石耐火材料是指以二氧化硅为主要化学成分的耐火材料，通常二氧化硅的含量大于93%。铝硅酸盐耐火材料简称铝硅系耐火材料，是指以氧化铝和二氧化硅为主要成分的耐火材料。其按氧化铝含量的不同又可以分为黏土质耐火材料（氧化铝含量大于或等于30%且小于45%）、高铝质耐火材料（氧化铝含量大于45%）等。镁质耐火材料是指氧化镁含量大于80%的耐火材料。镁尖晶石质耐火材料主要是由镁砂和氧化镁含量大于等于20%的尖晶石组成的耐火材料。镁铬质耐火材料是指由镁砂和铬铁矿制成的且以镁砂为主要组分的耐火材料。镁白云石质耐火材料是指由镁砂和白云石熟料制成的且以镁砂为主要组分的耐火材料。白云石质耐火材料是指以白云石熟料为主要原料的耐火材料。碳复合耐火材料也成为含碳耐火材料，是由氧化物、非氧化物以及石墨等碳素材料构成的复合材料。

按照耐火材料是否经过高温烧成可将耐火材料分为烧成耐火材料和不烧耐火材料。按照生产方式分类又可将耐火材料分为机压成型耐火材料、手工成型耐火材料、浇注耐火材料和熔铸耐火材料。

使用特殊的原料、采用特殊工艺制备或者具有特殊用途的耐火材料称为特种耐火材料，其在组成、生产工艺以及使用条件上不同于传统的耐火材料。

总之，耐火材料的分类方法取决于考虑的因素和用途，不同的分类方式有助于更好地理解和应用这些材料。本书以化学性质分类进行章节设置。另外，本书将特种耐火材料制品、不定形耐火材料制品分别设为一章，以便读者进行查阅。

2 酸性耐火材料制品

2.1 硅质耐火材料制品

2.1.1 术语词组

硅砖 silica brick
石灰结合硅质耐火材料 lime-bonded silica refractory
焦炉硅砖 coke oven silica bricks
热风炉硅砖 hot-blast stove silica bricks
高密度硅砖 high density silica bricks
轻质隔热硅砖 lightweight insulation silicon bricks

2.1.2 术语句子

硅砖中裂纹包括表面裂纹和层裂。

The crack occurred in **silica brick** generally includes surface crack and layer crack.

根据文献记载，第一块**硅砖**是 W. W. Young 在 1820 年制成的，他使用石灰作为硅石砂的黏合剂。

According to the literature, the first **silica brick** was made in 1820 by W. W. Young, who used lime as a bond for silica sand.

硅砖因具有优异的高温体积稳定性而被广泛应用于焦炉、热风炉和玻璃窑炉等领域。

Silica bricks are widely used in coke ovens, hot blast stoves and glass furnaces due to their excellent high temperature volume stability.

以碳酸钙作为**硅砖**的矿化剂时，砖内液相的生成温度提升至 1436 ℃，在此之前砖内发生的反应主要为碳酸钙的分解以及 β-石英至 α-石英至亚稳态 α-方石英的晶型转变。

When calcium carbonate is used as the mineralizer of **silica brick**, the formation temperature of liquid phase in the brick is increased to 1436 ℃. Before that, the reaction in the brick is mainly the decomposition of calcium carbonate and the transformation from β-quartz to α-quartz to metastable α-cristobalite.

高密度导热硅砖主要应用于焦炉砌筑燃烧室和碳化室的隔墙。

High density thermal conductive silica brick is mainly used in the partition wall of combustion chamber and carbonization chamber of coke oven.

在**硅砖**的制造中，以石英为原料，氧化钙被用作结合剂以及矿化剂促使在硅砖烧制过程中将方石英转化为鳞石英。

In manufacturing of **silica bricks**, quartzite has been used as a starting raw material and calcium oxide is used as a bonding material as well as a mineralizer to convert cristobalite into tridymite during the firing of green silica brick.

为保证石英充分转化为方石英和鳞石英，从而最大限度地减少使用过程中的永久膨胀，**硅砖**通常要烧至 1450 ℃左右，并在最高温度下保温相当长的时间。

To ensure adequate transformation of quartz to cristobalite and tridymite and consequent minimization of permanent expansion in service, **silica bricks** are usually fired to about 1450 ℃, with a fairly long soak at top temperature.

生产**致密硅砖**较为困难，主要是因为在烧制过程中原料向密度较低的形式转变，打乱了颗粒的初始排列。

The production of **dense silica bricks** is no means easy, mainly because the conversion of the raw material to a less dense form during firing upsets the original arrangement of the grains.

SiO_2 含量较高的**硅砖**抗碱侵蚀能力更强，并且以方石英、鳞石英和熔融石英为主晶相的**硅质耐火材料**都具有良好的抗碱侵蚀能力。

The **silica brick** with higher SiO_2 content has stronger alkali corrosion resistance, and the **siliceous refractory materials** with cristobalite, tridymite and fused quartz as the main crystal phases have good alkali corrosion resistance.

2.1.3 术语段落

1450 ℃使用前后**硅砖**的蠕变率分别为0.008%、-0.025%；1550 ℃使用前后硅砖的蠕变率分别为-0.154%、-0.135%。用后硅砖只有鳞石英相存在，1450 ℃用后硅砖蠕变与纯鳞石英陶瓷的蠕变特征相当。用后硅砖中鳞石英呈连续网络状分布，表现为优良的抗蠕变性能。1550 ℃用后硅砖蠕变的特征为部分鳞石英向方石英转变过程。硅砖微观结构从使用前的典型骨料-基质耐火材料结构演变为使用后的均质化鳞石英陶瓷材料结构[1]。

The creep rates of the **silica bricks** before and after use at 1450 ℃ are 0.008% and -0.025%, and the creep rates of the silica bricks before and after use at 1550 ℃ are -0.154% and -0.135%, respectively. The tridymite exists only in the silica bricks after use. The creep of the silica bricks is equivalent to the creep characteristic of the pure tridymite ceramic at 1450 ℃. The tridymite in the used bricks is distributed into a continuous network, showing an excellent creep resistance. The creep of the used bricks occurs due to a transformation process from partial tridymite to cristobalite at 1550 ℃. The microstructure of the silica bricks evolves from a typical aggregate-matrix refractory structure before use to a homogenized tridymite ceramic material structure after use[1].

研究了中国**硅砖**和硅质原料经高温热处理后的物相和形貌变化，以及硅砖的高温压蠕变行为。结果表明，二氧化硅原料在室温下呈致密结构，但热处理后出现裂纹和气孔。随着二氧化硅颗粒的破碎，硅砖转变为均质结构，结晶硅砖的最终均质度高于胶结硅砖。硅砖的最终蠕变结果显示在高温压蠕变测试（1550 ℃，50 h）后膨胀，并且由结晶二氧化硅制成的硅砖的膨胀更大。与国家标准GB/T 5073—2005相比，将L_0作为硅砖的初始高度，可以更全面地反映出硅砖的膨胀蠕变特性[2]。

The phase and morphology changes of both Chinese silica brick and silica raw material after high-temperature heat treatment, as well as the high-temperature compressive creep behaviour of **silica bricks** were investigated. The results show that silica raw materials presented a dense structure at room temperature, but cracks and pores appeared after heating treatment. With the crushing of silica particles, silica brick was transformed into homogeneous structure, and final homogeneous degree of the silica brick made of crystalline silica is higher than that of cemented silica. The final creep results of the silica bricks showed expansion after high-temperature compressive creep testing (1550 ℃, 50 h), and the expansion of silica brick made of crystalline silica was larger. Compared with the Chinese standard GB/T 5073—2005, setting L_0 as the original height of the silica brick can more fully reflect the expansion creep characteristics of silica brick[2].

目前，国内生产**硅砖**广泛采用CaO-FeO系矿化剂。其中，FeO主要以Fe_2O_3+FeO含量大于90%的轧钢皮（铁鳞）的形式引入；氧化钙因极易水化，难以在大气中保存，以石灰乳（氢氧化钙溶液）的形式引入。不同于国外单独使用钙质矿化剂生产硅砖，铁质矿化剂的引入主要是由于早期国内硅砖的烧成温度达不到要求：CaO-SiO_2二元系的最低共熔点为1436 ℃，而当时的硅砖烧成温度仅能达到1430 ℃左右。烧成温度不足导致当时生产的硅砖内残余石英含量难以控制，进而影响硅砖产品的膨胀、高温蠕变性能等多项性能指标。国内铁质矿化剂的引入在硅砖中构建FeO-Fe_2O_3-SiO_2三元系，将砖内的液相生成温度降低至最低1140 ℃，较低的液相生成温度使砖内的石英相能够在烧成过程中更加充分地通过溶解-析晶过程转变为鳞石英相，进而有效降低成品砖内的残余石英含量。硅砖的生产工艺发展至今，虽然天然气的广泛使用已经使得隧道窑的烧成温度能够达到1460 ℃，仍沿用石灰乳+铁鳞的矿化剂组合，目前的硅砖产品已经鲜有残余石英含量超标的现象发生[3]。

At present, CaO-FeO mineralizer is widely used in domestic production of **silica bricks**. Among them, FeO is mainly introduced in the form of rolled steel sheet (iron scale) containing Fe_2O_3+ FeO > 90 %. Calcium oxide is easily hydrated and difficult to preserve in the atmosphere. It is introduced in the form of lime milk (calcium hydroxide solution). Different from the production of silica bricks using calcium mineralizer alone abroad, the introduction of iron mineralizer in China is mainly due to the fact that the firing temperature of early domestic silica bricks cannot meet the requirements: the lowest eutectic point of CaO-SiO_2 binary system is 1436 ℃, while the firing temperature of silica bricks at that time can only reach about 1430 ℃. The insufficient firing temperature makes it difficult to control the residual quartz content in the silicon brick produced at that time, which in turn affects the expansion and high temperature creep properties of the silicon brick products. The FeO-Fe_2O_3-SiO_2 ternary system was constructed in the silica brick by the introduction of domestic iron mineralizer, which reduced the liquid phase formation temperature in the brick to a minimum of 1140 ℃. The lower liquid phase formation temperature enables the quartz phase in the brick to be more fully transformed into the tridymite phase through the dissolution-crystallization process during the firing process, thereby effectively reducing the residual quartz content in the finished brick. The production process of silica brick has been developed so far.

Although the wide use of natural gas has made the firing temperature of tunnel kiln reach 1460 ℃, the combination of lime milk and iron scale mineralizer is still used. At present, the content of residual quartz in silica brick products has rarely exceeded the standard[3].

以结晶二氧化硅为原料，氮化硅铁为矿化剂，制成新型**硅砖**。对添加1%、1.5%和2%氮化硅铁制备的新型硅砖的物相、显微结构、气孔特征和高温性能进行了表征。结果表明，与常规硅砖（矿化剂：氢氧化钙和铁鳞）相比，添加氮化硅铁的新型硅砖含有较少的氧化钙、较少的残余石英和较多的鳞石英。在硅砖的烧制阶段，氮化硅铁逐渐氧化形成 SiO_2 和 Fe_2SiO_4。加入氮化硅铁后，颗粒破碎程度降低，连通孔孔径增大。随着氮化硅铁含量的增加，硅砖结构逐渐致密化。含有2%氮化硅铁的硅砖样品在高温（1550 ℃，0.2 MPa）下获得了更好的抗蠕变性，但是通过使用氮化硅铁作为矿化剂时，砖的荷重软化温度略有降低[4]。

Novel **silica bricks** were made of crystalline silica with ferrosilicon nitride as the mineralizer. The phase, microstructure, pore characteristics and high-temperature properties of the novel silica bricks prepared by adding 1%, 1.5% and 2% ferrosilicon nitride were characterized. The results showed that, compared to conventional silica bricks (with mineralizers of calcium hydroxide and iron scale), the novel silica bricks with the addition of ferrosilicon nitride contained less calcium oxide, less residual quartz and more tridymite. The ferrosilicon nitride was gradually oxidized to form SiO_2 and Fe_2SiO_4 during the firing stage of the silica bricks. After adding ferrosilicon nitride, the degree of particle breakage was reduced, and the pore size of interconnected pores increased. With the increase in ferrosilicon nitride content, the silica brick structure gradually densified. The silica brick sample with 2% ferrosilicon nitride achieved better creep resistance at high temperature (1550 ℃, 0.2 MPa), but the refractoriness under load of the bricks was slightly reduced by using ferrosilicon nitride as a mineralizer[4].

选取了一些**硅质（95% SiO_2）耐火材料**，并表征它们的化学和结晶组成、显微结构、负荷下的耐火性和抗弯强度与温度的关系。蠕变试验是在高于1500 ℃的温度下使用差分方法实现的。这部分工作主要集中在一种材料上。补充测试：退火处理后进行晶体学分析，还进行了退火处理之后的晶体学分析、尺寸变化测量和微观结构观察。退火和蠕变试验的结果证明了该材料的收缩和体积膨胀之间的竞争，伴随着晶体组成和微观结构的转变。对结果进行了讨论，与文献进行了比较，并建立了一个模型，以解释这些行为，特别是膨胀效应。钙硅玻璃相的出现和微观结构的变化解释了高温行为和压缩蠕变。高温下的材料膨胀可以用氧化钙从基质迁移到聚集体来解释。冷却后的残余膨胀部分是由方石英相变引起的应力和开裂导致的[5]。

Some **silica (95% SiO_2) refractory materials** were selected and characterized by their chemical and crystallographic compositions, microstructure, refractoriness under load and bending strength versus temperature. Creep tests were realized at temperature higher than 1500 ℃, using the differential method. This part of the work was mainly focalized on one material. Complementary tests: annealing treatment followed by crystallographic analysis, measurement of dimensions variation and microstructure observation were also realized. Results of annealing and creep tests demonstrate a competition between shrinkage and a volume expansion of the material, associated with a modification of crystallographic composition and of the microstructure. Results were discussed, compared with literature and a model was developed to explain the behavior and especially the expansion. The high temperature behavior and the compressive creep were explained by the appearance of a calcia-silica vitreous phase and microstructure change. The material expansion at high temperature is explained by calcia migration, from matrix into aggregates. Residual expansion after cooling is partially explained by the cristobalite transformation which induces stress and cracking[5].

2.2 半硅质耐火材料制品

2.2.1 术语词组

半硅质耐火材料 semi-silica refractories
半硅质原料 semi-silica raw materials
焦炉用半硅砖 semi-silica brick for coke oven
叶蜡石砖 pyrophyllite bricks

2.2.2 术语句子

半硅质耐火材料是一种二氧化硅（化学组成）含量大于65%的耐火材料。

Semi-silica refractories are one kind of refractory with a minimum of 65% silica content as determined by chemical analysis.

半硅砖是另一种硅质耐火砖。这种耐火材料介于耐火黏土和硅质耐火砖之间。

Semi-silica bricks are another type of silica refractory bricks. This type of refractory is an intermediate between fire clay and silica refractory bricks.

半硅砖在与熔渣接触时具有上釉倾向，当在类似条件下硅砖或耐火黏土砖被完全渗透时，而半硅砖侵蚀仅限于表面。

Semi-silica bricks possess an tendency to glaze in contact with slags, attack being limited to the surface when under similar conditions silica or fireclay bricks are fully penetrated.

我国**叶蜡石砖**是20世纪70年代初研制成功的，近年来有较快的发展。

Pyrophyllite bricks in China were successfully developed in the early 1970s and have developed rapidly in recent years.

叶蜡石砖是以天然叶蜡石为原料生产的耐火制品，而天然叶蜡石是由叶蜡石（$Al_2O_3 \cdot 4SiO_2 \cdot H_2O$）和含水铝硅酸盐矿物组成。

Pyrophyllite brick is a refractory product produced from natural pyrophyllite, which is composed of pyrophyllite（$Al_2O_3 \cdot 4SiO_2 \cdot H_2O$）and hydrous aluminosilicate minerals.

与传统的硅酸铝质耐火材料生产工艺相比，**叶蜡石砖**具有生料制砖、高压成型、低温烧成等工艺特点。

Compared with the traditional aluminum silicate refractory production process, **pyrophyllite brick** has the characteristics of raw material brick making, high pressure forming and low temperature firing.

2.2.3 术语段落

由天然原料或人造砂黏土混合物制成，SiO_2 含量为70%~93%的**半硅砖**仍被广泛应用，例如在蓄热室和加热炉炉顶的下层。除了相对便宜之外，它们比普通的黏土砖体积更稳定，比硅石具有更好的抗热震性[6]。

Semi-silica bricks containing 70%-93% SiO_2 made from natural or artificial sand-clay mixtures still have a useful role to play, e. g. in the lower courses of regenerators and reheating-furnace roofs. Apart from their relative cheapness they are more volume stable than common fireclay bricks and have a higher thermal-shock resistance than silica[6].

叶蜡石砖在加热过程中有相变，并伴随有体积变化效应，这一特性与硅砖相近，与黏土砖、高铝砖和镁砖有根本区别。但叶蜡石砖的烧成目的与硅砖又完全不同，硅砖在于使石英充分转化，硅砖烧成的好坏取决于硅石转变的程度和转化的趋向，也就是真比重的高低、残余膨胀的大小和鳞石英化的多少。而叶蜡石砖的烧制目的是在保证制品强度和密度的情况下，尽量延缓石英的方石英化。也就是说，烧制过程中必须防止石英向方石英转化，使制品尽量保持更多的残存石英，减少线膨胀[6]。

Pyrophyllite bricks have phase transition during heating, accompanied by volume change effect, which is similar to silica bricks, and fundamentally different from clay bricks, high alumina bricks, and magnesia bricks. However, the firing purpose of the wax brick is completely different from that of the silica brick. The silica brick is to make the quartz fully transformed. The firing quality of the silica brick is determined by the degree of silica transformation and the trend of transformation, that is, the true specific gravity, the size of the residual expansion and the amount of scaly quartz. The purpose of burning wax brick is to delay the cristobalite of quartz as much as possible while ensuring the strength and density of the product. That is to say, the conversion of quartz to cristobalite must be prevented during the firing process, so that the product can keep as much residual quartz as possible and reduce linear expansion[6].

叶蜡石砖保存大量的残存石英，杂质成分很少，其制品的荷重软化温度也比较高。因为高温下液相少，而且液相黏度大，与渣接触反应少。通过叶蜡石砖荷重软化温度的技术指标，可以推测叶蜡石砖的使用效果，特别是与渣接触仍能保持较高的荷重软化温度，制品才能有较高的使用寿命。如果碱含量增加，其荷重软化温度有所降低[7]。

Pyrophyllite bricks preserve a large amount of residual quartz, few impurity components, and the load softening temperature of its products is also relatively high. Because the liquid phase is less at high temperature, and the liquid phase viscosity is large, the contact reaction with the slag is less. Through the technical index of the load softening temperature of the wax stone brick, the use effect of the wax stone brick can be inferred, especially the contact with the slag can still maintain a high load softening temperature, and the product can have a high service life. If the alkali content increases, the load softening temperature decreases[7].

为了减少环境污染，利用褐煤粉煤灰废料作为制备**半硅质隔热耐火砖**的主要原料。燃煤火力发电厂产生大量的褐煤飞灰作为废物。这些废料含有对环境有害的有毒元素。本研究以褐煤粉煤灰废料为主要原料制备半硅质隔热耐火砖，以减少环境污染。褐煤灰、球黏土和锯末是半硅质隔热耐火砖的主要成分。粉煤灰和球黏土以不同比例混合，木屑用作耐火砖中的造孔剂。制备的样品在1000~1200 ℃范围内烧制 2 h。这些砖的最终性能通过机械、热、物相分析和显微结构来表征。发现粉煤灰可作为制备半硅质隔热耐火砖的合适原材料。用褐煤粉煤灰制备的隔热耐火砖的化学分析、体积密度（BD）、显气孔率（AP）和导热系数（TC）与市售产品相似。含 60%褐煤粉煤灰、30%球黏土和 10%锯末的 FC30S 样品在1100 ℃烧结 2 h 后，显气孔率为 44.69%，导热系数为 0.38 W/(m·K)。因此，粉煤灰的使用在技术上是可行的，经济上是有利的，可用作隔热耐火材料。该研究为利用粉煤灰制备环保型半硅质隔热耐火材料指明了一条可持续发展的技术途径[8]。

The lignite fly ash waste materials are used as the primary raw material for the fabrication of **semi-silica insulation refractory bricks** to reduce the environmental pollution. A tremendous amount of lignite fly ash (FA) is produced as waste materials in coal-fired thermal power plants. These wastes materials contain toxic elements which are detrimental to the environment. In this study, the lignite FA waste materials are used as the primary raw material for the fabrication of semi-silica insulation refractory bricks to reduce the environmental pollutions. Lignite FA, ball clay, and sawdust are the main components of semi-silica insulation refractory bricks. FA and ball clay were blended in various combinations, and sawdust is used as a pore former in the refractory bricks. The prepared samples were fired in the range of 1000 ℃ to 1200 ℃ for 2 h. The resultant properties of these bricks were characterized in terms of mechanical, thermal, phase analysis, and microstructure. It was found that FA can be used as a suitable candidate in the preparation of semi-silica insulation refractory bricks. The chemical analysis, bulk density (BD), apparent porosity (AP), and thermal conductivity (TC) of the insulation refractory bricks prepared with lignite FA were observed to be similar to the commercial products. The sample FC30S, which contains 60% lignite FA, 30% ball clay, and 10% sawdust showed 44.69% AP and 0.38 W/(m·K) thermal conductivity after sintering at 1100 ℃ for 2 h. Hence, the use of FA is technically feasible, economically beneficial, and can be used as thermal insulation refractory. The study indicates a sustainable technique of using fly ash for the fabrication of eco-friendly semi-silica insulation refractory[8].

2.3 黏土耐火材料制品

2.3.1 术语词组

黏土砖 clay bricks
普通耐火黏土砖 common fireclay refractory bricks
致密黏土砖 dense fireclay bricks
优质耐火黏土砖 high-duty fireclay bricks
低气孔耐火黏土砖 low porosity fireclay bricks
磷酸浸渍耐火黏土砖 phosphoric acid immersed fireclay bricks
多熟料耐火黏土砖 high grog fireclay bricks
轻质隔热黏土砖 lightweight insulating fireclay bricks
耐火黏土绝热板 fireclay insulating board
黏土袖砖 fireclay sleeve
黏土格子砖 fireclay checker bricks
耐火黏土 fire clay

2.3.2 术语句子

耐火黏土是人类使用的第一种耐火材料。

Fireclay was probably the first refractory used by man.

耐火黏土砖的制造方法有两种，一种是干压成型，另一种是挤压黏土柱，然后再压制成所需的形状。

Fireclay bricks are manufactured either by dry pressing or by extruding a column of clay, followed by repressing to give the desired shape.

黏土质耐火材料是由天然黏土（富 SiO_2）制成的，它可以承受高于 PCE（高温圆锥当量）值 19 的温度，且不会分解、变形、破裂、软化或熔化。

Fireclay refractories are formed from natural clay (silica-rich), which can withstand a temperature above the PCE (pyrometric cone equivalent) value of 19 without disintegrating, deforming, cracking, softening, or melting.

上层格子砖在使用中被焦油浸泡，使得其易于黏结煤气中的粉尘，导致与**黏土砖**的反应。

The upper lattice brick is soaked in tar during use, which makes it easy to bond the dust in the gas and cause reaction with the **clay brick**.

黏土砖发泡现象除与黏土格子砖的杂质含量和性能有关外，也与煤气中的粉尘有关。

The foaming of **clay brick** is not only related to the impurity content and performance of clay lattice brick, but also to the dust in the gas.

2.3.3 术语段落

低气孔黏土砖主要应用于玻璃熔窑蓄热室的下部，作为格子砖或墙体。它们在高温下承受上部耐火材料的载荷。因此，在预测回热器的寿命时，应考虑它们的蠕变行为。选择两种低气孔黏土砖作为试验样品。结果表明，它们在不同压应力（0.1 MPa、0.2 MPa 和 0.3 MPa）和不同温度（1200 ℃和 1280 ℃）下的蠕变速率在 0.2%以内。具有较低显气孔率的 LPFB1 试样对温度更敏感。LPFB2 试样对压力更敏感。蠕变与其他力学性能的关系为：Al_2O_3 含量越高，体积密度越高，显气孔率越低，低气孔黏土砖的荷重软化温度越高，总应变范围越小[9]。

Low porosity fireclay bricks (LPFBs) are mainly applied in the lower party of glass furnace regenerator as checker works or walls. They are subjected to the loads of upper refractory materials at high temperature. So the creep behavior of them should be considered to predict the longevity of the regenerator. Two kinds of LPFBs were chosen as the test samples. It is shown that the creep rates of them under different stresses (0.1 MPa, 0.2 MPa, and 0.3 MPa) and different temperatures (1200 ℃ and 1280 ℃) are within 0.2%. LPFB1 with lower apparent porosity is more sensitive to temperature. LPFB2 is more sensitive to stress. The relationship between the creep and other mechanical properties is: the higher the Al_2O_3 content, the higher the bulk density, the lower the apparent porosity, and the higher the refractoriness under load of LPFB2, the lower the total strain range of it[9].

具体而言，在用废耐火砖制**烧结黏土砖**时，抗压强度和吸水率是两个重要指标。生产烧结砖中，若以抗压强度为考虑重点时，不论在氧化气氛还是还原气氛下焙烧，废粒料使用粒径在 7 mm 以下的，配合比设定在 50%（按体积计）以下。即若废粒料粒径在 3～7 mm 范围内，其配合比在 20%以下（按体积计）为好；若其粒径在 1～3 mm 范围内，配合比在 40%以下（按体积计）为好；若其粒径在 1 mm 以下，配合比在 50%以下（按体积计）为好。若以吸水率为考虑重点，不论氧化还是还原焙烧，所用废粒料粒径在 7 mm 以下，其配合比最好设定在 10%以上（按体积计）。即不论其粒径在 3～7 mm、1～3 mm，还是在 1 mm 以下，其配合比均以30%以上（按体积计）为好。若以抗压强度和吸水率同时重点考虑，所用废粒料的粒径在7 mm以下，而其配合比以10%~50%（按体积计）为好，即其粒径在1~3 mm，配合比设定在30%~40%（按体积计）为好，其粒径在1 mm以下，配合比设定在30%~50%（按体积计）为好。总之，废粒料与黏土配合比依据烧结砖主要性能即抗压强度、吸水率和颗粒不同粒径来适当设定[10]。

Specifically, compressive strength and water absorption are two important indicators when **sintering clay bricks** with waste refractory bricks. In the production of sintered bricks, if the compressive strength is taken as the focus, whether it is calcined in oxidizing atmosphere or reducing atmosphere, the particle size of waste pellets is less than 7 mm, and the mix ratio is set below 50% (according to volume). That is, if the particle size of waste particles is in the range of 3-7 mm, the mixture ratio is better than 20% (by volume); if the particle size is in the range of 1-3 mm, the mixture ratio is better than 40% (according to volume); if the particle size is below 1 mm, the mixture ratio is below 50% (by volume). If the water absorption is taken as the focus, regardless of oxidation or reduction roasting, the particle size of the waste pellets used is below 7 mm, and the mix ratio is preferably set above 10% (calculated by volume). That is, whether its particle size is 3-7 mm, 1-3 mm, or below 1 mm, its mix ratio is better than 30% (according to volume). If the compressive strength and water absorption are considered at the same time, the particle size of the waste pellets used is below 7 mm, and the mix ratio is 10%-50% (according to volume), that is, the particle size is 1-3 mm, and the mix ratio is set at 30%-40% (according to volume). It is good that the particle size is below 1mm and the mix ratio is set at 30%-50% (according to volume). In short, the mixture ratio of waste particles and clay is appropriately set according to the main properties of sintered bricks, namely compressive strength and water absorption and different particle sizes of powders[10].

大型焦炉蓄热室多为**黏土格子砖**与半硅质格子砖分层砌筑，也有全部采用黏土格子砖砌筑。近年来，有的焦炉因为材质问题导致检修周期缩短，有的焦炉黏土格子砖出现发泡、软熔变形堵塞格栅孔道，分析原因往往归结于熔渣的渗透破坏，但是熔渣从何而来？究竟是什么因素导致格子砖的软熔变形？文献中都没有介绍。焦炉黏土格子砖KA40的技术条件要求为：$w(Fe_2O_3) \leqslant 2\%$，$w(K_2O+Na_2O) \leqslant 1\%$，$w(K_2O+Na_2O+CaO+MgO) \leqslant 1.8\%$；使用前的新砖如果检验合格，其成分应该满足技术条件的要求；而从文献所分析的黏土砖新砖结构看，骨料和基质都不存在含铁氧化物的富集区域，那么用后黏土砖的含铁氧化物从何而来？结构分析中的含铁氧化物的富集区域是如何产生的？这些问题是涉及焦炉蓄热室长寿与选材的关键。如果是黏土砖的质量问题，只需要提高其质量就可以延长蓄热室格子砖的使用寿命；如果是工艺问题，格子砖更换再好仍然存在发泡软熔变形的隐患。黏土格子砖的发泡变形不仅存在于7.63 m焦炉，也存在于7 m焦炉的炉役后期，更换格子砖工程量大，也增加成本，影响生产效率[11]。

Most of the large coke oven regenerators are layered masonry of **clay lattice bricks** and semi-silica lattice bricks, and all of them are masonry of clay lattice bricks. In recent years, some coke ovens have shortened the maintenance cycle due to material problems. Some coke oven clay lattice bricks have foaming and softening deformation to block the grid pores. The analysis reasons are often attributed to the penetration damage of slag, but where does the slag come from? What factors lead to the soft melting deformation of lattice bricks? It is not introduced in the literature. The technical requirements of KA40 are : $w(Fe_2O_3) \leq 2\%$, $w(K_2O+Na_2O) \leq 1\%$, $w(K_2O+Na_2O+CaO+MgO) \leq 1.8\%$; if the new brick before use is qualified, its composition should meet the requirements of technical conditions; from the new brick structure of clay brick analyzed in the literature, there is no enrichment area of iron oxide in aggregate and matrix, so where does the iron oxide of clay brick come from? How is the enrichment region of iron oxides in the structural analysis generated? These problems are the key to the longevity and material selection of coke oven regenerator. If it is the quality problem of clay brick, only need to improve its quality can prolong the service life of the regenerator lattice brick; if it is a process problem, the lattice brick replacement is still a hidden danger of foaming softening deformation. The foaming deformation of clay checker brick not only exists in 7.63 m coke oven, but also in the later stage of 7 m coke oven. The replacement of checker brick has a large amount of work, which also increases the cost and affects the production efficiency[11].

莫来石（$3Al_2O_3 \cdot 2SiO_2$）具有抗蠕变能力高、热膨胀系数和导热系数低、较好的抗热震性能和抗腐蚀性能好等优异的性能，是一种很有前景的高温结构耐火材料。而莫来石晶须不但具有莫来石质材料的性能，而且作为长径比较大的完整单晶体具有更加优良的力学性能，其优异的高温抗蠕变性能使其特别适合作为高温环境用耐火材料的增强增韧晶须使用，是一种优异的陶瓷基复合材料和金属基复合材料增强增韧相。由于合成的莫来石晶须和基体材料与高铝黏土砖（主要成分：刚玉-莫来石）具有共同的成分，因此，探讨合理的工艺路线，通过添加适当的添加剂原位合成莫来石晶须，进而通过晶须的拔出、裂纹的桥接等机制起到增强增韧高铝黏土砖耐火材料的作用，这不仅可以提高**高铝黏土砖耐火材料**的抗热冲击性能和使用寿命，还有助于降低耐火材料的制造成本[12]。

Mullite ($3Al_2O_3 \cdot 2SiO_2$) has excellent properties such as high creep resistance, low thermal expansion coefficient and thermal conductivity, good thermal shock resistance and corrosion resistance. It is a promising high-temperature structural refractories. Mullite whiskers not only have the properties of mullite materials, but also have better mechanical properties as a complete single crystal with a large aspect ratio. Its excellent high temperature creep resistance makes it particularly suitable for strengthening and toughening whiskers as refractory materials for high temperature environments. It is an excellent ceramic matrix composite and metal matrix composite reinforcement and toughening phase. Since the synthesized mullite whiskers and the matrix material high-alumina clay brick (main component: corundum-mullite) have a common composition, a reasonable process route is explored to synthesize mullite whiskers in situ by adding appropriate additives, and then the whiskers are pulled out. The bridging of cracks and other mechanisms play a role in strengthening and toughening high-alumina clay brick refractory materials, which can not only improve the thermal shock resistance and service life of **high-alumina clay brick refractory materials**, but also help to reduce the manufacturing cost of refractory materials[12].

添加0%、10%、20%和30%的氧化铝磨屑废料制备**耐火黏土砖**。以150 MPa的单轴压制制备样品，并在不同的温度900~1100 ℃下烧制。对制备的含有和不含氧化铝磨屑废料耐火黏土砖样品的物理性质如体积密度、显气孔率、线性收缩率和吸水率进行测定。结果表明，BFC30 黏土砖的强度和体积密度有所提高。BFC30 的最大抗压强度为31. 15 MPa，吸水率为5. 84%，线收缩率为1. 64%。在BFC30 样品中观察到最高的抗弯强度为35. 34 MPa。此外，增加氧化铝磨屑废料质量分数进而增加耐火黏土砖中的氧化铝和二氧化硅含量来改善力学性能。在最低撞击角为30°时，BFC30 样品的体积磨损率最低为2. 4 g/mm^3[13]。

The present work deals with the production of **fireclay brick** by adding different compositions of 0%, 10%, 20% and, 30% of alumina abrasive waste. Samples are prepared by uniaxial pressing of 150 MPa and fired at different temperatures such as 900 ℃ and 1100 ℃. The physical properties such as bulk density, apparent porosity, linear shrinkage, and water absorption were estimated for the prepared fireclay brick samples with and without alumina abrasive waste. The results show that fireclay brick samples of BFC30 possess an increase in strength and bulk density. BFC30 formulation exhibited the maximum compressive strength of 31. 15 MPa, with lower water absorption of 5. 84% and linear shrinkage value of 1. 64%. The highest bending strength value of 35. 34 MPa was observed in the BFC30 sample. Furthermore, the increase in alumina abrasive waste mass fraction improves the mechanical properties by increasing the alumina and silicon dioxide content in the produced fireclay brick. The lowest volumetric wear rate of 2.4 g/mm^3 was observed on the BFC30 sample at the lowest impingent angle of 30°[13].

2.4 高铝质耐火材料制品

2.4.1 术语词组

高铝砖 high alumina bricks
Ⅲ等高铝砖 grade Ⅲ high alumina bricks
Ⅱ等高铝砖 grade Ⅱ high alumina bricks
Ⅰ等高铝砖 grade Ⅰ high alumina bricks
高铝质耐火材料 high-alumina refractory
高铝耐火砖 high-alumina refractory bricks
高铝质制品 high alumina products
不烧高铝砖 unfired high alumina bricks
氧化铝耐火砖 alumina firebricks
高铝塞头砖 high-alumina stopper
抗剥落高铝砖 spalling resistant high-alumina bricks
低蠕变高铝砖 low creep high alumina bricks
抗蠕变高铝砖 creep-resistant high-alumina bricks
矾土基高铝砖 bauxite-based high alumina bricks
盛钢桶用高铝砖 high alumina bricks for casting ladle
矾土耐火材料 bauxite refractories

2.4.2 术语句子

水泥窑用**抗剥落高铝砖**。
Spalling resistant high alumina bricks for cement kiln.

高炉用**抗蠕变高铝砖**。
Creep resistant high-alumina brick for blase furnaces.

Ⅱ等高铝砖具有更高的强度值，在断裂前可以承受更大的应变。
Grade Ⅱ high alumina brick possesses higher strength values and can accommodate more strain prior to fracture.

高铝质耐火材料的高温力学性能与相组成和显微结构特征有关。

The high temperature mechanical behaviour of **high alumina refractories** may be correlated with phase composition and microstructural characteristics.

矾土基耐火材料的高温力学性能与其 Al_2O_3 含量密切相关。Al_2O_3 含量越接近 70%，高温力学性能越好。

The high temperature mechanical properties of **bauxite refractories** are closely associated with the Al_2O_3 content. The closer is the Al_2O_3 content to 70%, the better the high temperature mechanical properties.

Ⅲ等高铝砖抗蠕变性较差的原因是高玻璃含量（15%~25%）高和缺少晶体间的结合。

The inferior creep resistance of **grade Ⅲ high alumina brick** may be interpreted in terms of its high glass content (15%-25%) and its lack of crystal-to-crystal bonding.

为满足高炉炉衬在更高风温下操作的应用需求，以烧结矾土为主要原料，开发了一系列**矾土基抗蠕变高铝砖**。

For application in blast furnace stoves operating at higher blast temperatures, a series of **bauxite-based creep resistant high alumina bricks** has been developed, using sintered bauxite as the main staring material.

高铝矾土是氧化铝的主要来源，氧化铝在耐火材料工业中用于制备各种有价值的耐火材料产品[13]。

Bauxite is the main source of alumina which is used in refractory industries for the preparation of various valuable refractory products[13].

这些结果表明，含 ZrO_2 的原料在保温性能方面优于铝土矿基材料。**含 ZrO_2 铝矾土耐火材料**的耐磨性良好，是水泥工业低碳生产中很有前景的耐火材料。

These results suggest that ZrO_2-containing raw materials outperform bauxite-based materials with respect to the thermal insulation performance. **ZrO_2-containing bauxite refractories** can be promising refractory materials for low-carbon production of cement industry as long as their abrasion resistance is satisfactory.

2.4.3 术语段落

采用研制的高温弯曲应力-应变测试仪研究了 **Ⅰ等高铝砖**（GL-80）、**Ⅱ等高**

铝砖（GL-75）和Ⅲ等**高铝砖**（GL-55）在不同温度下的力学性能，包括应力-应变曲线、弹性模量（MOE）、抗折强度（MOR）和断裂时的最大变形量。结果表明：高铝砖在不同温度下的应力-应变关系可以分为弹性阶段、塑性阶段和黏滞流动阶段；在低、中温范围内，高铝砖的弹性模量和抗折强度随温度的上升而增加，到达某一转折温度后，随温度的上升而明显下降；3 种**高铝砖**高温力学性能从高到低的排列顺序为：Ⅱ等 > Ⅰ等 > Ⅲ等[14]。

The mechanical properties, such as stress-strain curves, modulus of elastic (MOE), modulus of rupture (MOR), maximum fracture deformation of high alumina bricks including **Grade** Ⅰ (DL-80), **Grade** Ⅱ (GL-75) and **Grade** Ⅲ (GL-55) were studied using the newly-developed hot bending stress-strain tester. The results show that the stress-strain relationship of the bricks at elevated temperatures may be divided into three stages, viz: elastic, plastic and viscous flow stages. MOR and MOE of the bricks tend to increase with the rising of temperature at low and medium temperatures up to an inflexion point, after which they decrease obviously. The order of merit for the three **high alumina bricks** in term of mechanical properties at elevated temperatures is grade Ⅱ > grade Ⅰ > grade Ⅲ[14].

为了解蠕变引起炉衬材料破坏的机理，对炉衬材料蠕变的三个阶段进行评价具有重要意义。本文利用高温蠕变试验装置，获得了**高铝砖**在 1350 ℃时 3 MPa、3.25 MPa 和 3.5 MPa 的三阶段蠕变曲线。然后采用 K-R 模型、Sinh 模型和 Liu-Murakami 模型结合 Garofalo 公式对实测蠕变曲线进行拟合。结果表明，3 种模型均适用于高铝砖蠕变行为的描述和蠕变寿命的预测，其中 Sinh-Garofalo 模型的拟合效果最好。因此，选择 Sinh-Garofalo 模型进行有限元建模。通过比较模拟值和解析值，验证了模型的可靠性。最后，通过对比考虑和不考虑蠕变损伤的模型，发现蠕变损伤模型可以用来表征蠕变损伤演化过程和预测结构的寿命，可进一步应用于炉衬的失效研究[15]。

In order to understand the mechanism of lining failure caused by creep, it is significant to evaluate the three creep stages of lining materials. In this paper, the three stages creep curves of 3 MPa, 3.25 MPa and 3.5 MPa for **high alumina bricks** at 1350 ℃ were obtained by high temperature creep testing equipment. Then the K-R model, the Sinh model and the Liu-Murakami model combined with the Garofalo formula are used to fit the measured creep curves. The results show that all the three models are suitable for the description of the creep behavior and the creep life prediction of high alumina bricks, and the Sinh-Garofalo model has the best fitting effect. Therefore, the Sinh-Garofalo model is selected for finite element modeling. The reliability of the model is verified by comparing the simulated and analytical values. Finally, by comparing the model with/without considering the creep damage, it is found that the creep damage model can be used to characterize the creep damage evolution process and predict the life of structures, which can be further used in the failure study of furnace lining[15].

本研究采用XRD、TG-DSC、SEM、粒度分布和流变学等测试手段，对累托石黏土和球黏土结合**不烧高铝砖**的结合机理和常温/高温性能进行了研究。结果表明，黏土颗粒在水中由于水化膨胀和静电排斥力而分离成层状结构单元，累托石层状结构单元比球形黏土具有更大的长径比。累托石层结构单元干燥后形成“面对面”的带状结构，黏结性能优于球黏土（边对面的纸牌屋结构）。在脱羟基温度下煅烧3 h后，9%累托石/球黏土结合不烧高铝砖的常温耐压强度分别达到71 MPa和50 MPa，可以满足大多数高铝砖运输和使用的强度要求。球黏土结合不烧高铝砖在高温下的二次莫来石化和较低的液相含量使其具有比累托石黏土结合砖更高的荷重软化温度和更低的线收缩率。9%累托石/球黏土结合不烧高铝砖的荷重软化温度$T_{0.6}$分别为1262.6 ℃和1580.3 ℃。因此，9%球黏土结合不烧高铝砖比9%累托石结合砖具有更广泛的使用温度范围[16]。

In this study, the bonding mechanism and normal/high temperature performance of rectorite clay or ball clay bonded **unfired high alumina bricks** were investigated by using different techniques (XRD, TG-DSC, SEM, particle size distribution and rheology). The results showed that clay particles are separated into layer structural units in water due to the hydration swelling and electrostatic repulsive force, and rectorite layer structural units have larger aspect ratio than ball clay. Rectorite layer structural units form band-type structure with "face to face" after drying results in better bonding performance than ball clay (card-house structure with "edge to face"). The cold crushing strength of 9% rectorite/ball clay bonded unfired high alumina bricks after firing at the dehydroxylation temperature for 3 h reach 71 MPa and 50 MPa, respectively, and which can satisfy the strength requirement for the transportation and use of most high alumina bricks. The secondary mullitization and lower liquid phase content of ball clay bonded unfired high alumina brick under high temperature cause it has higher refractoriness under load and lower linear shrinkage than rectorite clay bonded brick. The $T_{0.6}$ refractoriness under load of 9% rectorite/ball clay bonded unfired high alumina brick are 1262.6 ℃ and 1580.3 ℃, respectively. Thus, the 9% ball clay bonded unfired high alumina bricks have wider service temperature range than 9% rectorite bonded bricks[16].

由于过渡的二次膨胀效应，一些煅烧铝矾土在**高铝耐火材料**的使用中受到限制。将煅烧矾土与纯α-氧化铝进行比较，以确定引起体积膨胀的结晶变化，确定膨胀的机理，并提出控制膨胀的方法。使用X射线衍射和岩相技术，发现体积膨胀是由于杂质有助于氧化铝和二氧化硅的扩散，促进了二次莫来石结晶的结果。二次莫来石化反应和莫来石晶体生长在1400 ℃开始，并在1600 ℃引起多达23%的膨胀。膨胀主要是由莫来石生长引起的，而不是由莫来石化反应导致的真密度变化引起的。少量氟化钠的引入证实对控制体积膨胀是有效的[17]。

The use of some calcined bauxites in **high-alumina refractories** is limited by unusual and excessive secondary expansion. One such calcined bauxite was compared with pure alpha alumina to determine the crystalline changes responsible for the expansion, to determine the mechanism of the expansion, and to suggest means for control of the expansion. X-ray diffraction and petrographic techniques were used. The expansion was found to result from crystallization of secondary mullite by the action of impurities in aiding the diffusion of alumina and silica. Secondary mullitization and growth began at 1400 ℃. and caused as much as 23% expansion at 1600℃. The expansion was caused primarily by the mullite growth rather than by the true density changes associated with mullitization. Small sodium fluoride additions proved to be effective for controlling expansion[17].

高铝矾土熟料采用块状料煅烧而成，可分为多个等级。此外，相同氧化铝含量的矾土熟料根据煅烧方式又可以分为回转窑料、倒焰窑料、燃气竖窑料、土竖窑料等。煅烧致密程度大小依次为倒焰窑料（或连体窑料）>燃气竖窑和回转窑料>土竖窑料。85、88等高品级矾土熟料主要通过回转窑和倒焰窑煅烧制备[18]。

High bauxite clinker is calcined by block material, which can be divided into multiple grades. In addition, bauxite clinker with the same alumina content can be divided into rotary kiln material, inverted flame kiln material, gas shaft kiln material, soil shaft kiln material and so on according to the calcination method. The order of calcination density is inverted flame kiln material (or connected kiln material) > gas shaft kiln and rotary kiln material > soil shaft kiln material. High-grade bauxite clinkers such as 85 and 88 are mainly prepared by calcination in rotary kiln and inverted flame kiln[18].

在烧成**Al_2O_3-C滑板材料**配方中加入预合成的矾土基β-Sialon，制备了低碳Al_2O_3-Sialon滑板材料。其碳含量降低了3%~5%，高温抗折强度和抗热震性保持原有水平，而抗氧化性有明显提高。这种滑板材料已在70 t和160 t钢包试用，连续使用3~5炉次，氧化层厚度2 mm，比 Al_2O_3-C 材料的6 mm要小；扩孔2~3 mm/次，与 Al_2O_3-C 材料相当。这种材料有望更多用于连铸功能材料[18]。

Low carbon Al_2O_3-Sialon sliding plate material was prepared by adding pre-synthesized bauxite based β-Sialon to the formula of fired **Al_2O_3-C sliding plate material**. Its carbon content is reduced by 3%-5%, the high temperature folding strength and thermal shock resistance remain the original level, and the oxidation resistance has been significantly improved. This slide material has been tested in 70 t and 160 t ladle, continuous use of 3-5 times, oxide layer thickness of 2 mm, than the Al_2O_3-C material of 6 mm is smaller; Reaming 2-3 mm/times, similar to Al_2O_3-C material. This material is expected to be more used in continuous casting functional materials[18].

自主研发的**矾土基电熔锆刚玉**和尖晶石是用高铝矾土取代氧化铝，采用“二步还原熔炼，一步氧化精炼”创新工艺研制的。它的理化性能与氧化铝基材料相当，而成本可降低20%~30%，经过试制试用后，已经批量生产。电熔锆刚玉用于制备 Al_2O_3-ZrO_2-C 滑板，电熔尖晶石主要用于制备钢包铝镁衬砖和浇注料，都已取得良好的使用效果，应用前景很好[18]。

Independent research and development of **bauxite based fused zirconium corundum** and spinel is to replace alumina with high bauxite, using " two steps reduction melting, one step oxidation refining" innovation process. Its physical and chemical properties are comparable to alumina-based materials, but the cost can be reduced by 20% to 30%, after trial production and trial, has been mass produced. Fused zirconium corundum is used to prepare Al_2O_3-ZrO_2-C skateboard, fused spinel is mainly used to prepare aluminum-magnesium lining bricks and castable of ladle, all of which have achieved good results and have good application prospects[18].

矾土基耐火材料是指以高铝矾土为原料制备的烧结矾土熟料、高铝砖和高铝不定形材料。我国矾土基耐火材料从20世纪50年代开始即以较高速度发展并逐渐在高炉、热风炉、电炉炉顶、钢包内衬、加热炉和水泥窑等广泛使用，取得良好效果，为钢铁工业和其他高温工业做出显著贡献。与此同时，它们的出口量逐年有所增加，近几年每年出口高铝矾土熟料150万~200万吨，高铝制品15万~20万吨；估计我国高铝矾土熟料在国际市场约占60%的份额，已成为主要供应来源[19]。

Bauxite based refractory refers to sintered bauxite clinker, high alumina brick and high alumina amorphous material prepared with high alumina as raw material.

China's bauxite based refractory materials have been developed at a high speed since the 1950s and gradually used in blast furnaces, hot blast furnaces, electric furnace tops, ladle linings, heating furnaces and cement kilns, achieving good results, and making significant contributions to the iron and steel industry and other high-temperature industries. At the same time, their exports have increased year by year, in recent years, the annual export of high bauxite clinker 1.5 million to 2 million tons, high aluminum products 150000 to 200000 t; It is estimated that China's bauxite clinker accounts for about 60% of the international market and has become the main source of supply[19].

2.5 莫来石耐火材料制品

2.5.1 术语词组

莫来石砖 mullite bricks
刚玉莫来石砖 corundun-mullite bricks
硅线石砖 sillimanite bricks
莫来石陶瓷 mullite ceramics
莫来石熟料 mullite chamotte
矾土基莫来石 bauxite-based mullite
矾土基莫来石均质料 bauxite-based homogenized mullite grogs
烧结莫来石制品 sintered mullite products
熔融莫来石 fused mullite
莫来石铸块 mullite blocks
莫来石刚玉制品 mullite-corundum products
莫来石刚玉质耐火材料 mullite-corundum refractory
莫来石刚玉格子砖 mullite-corundum checker bricks
莫来石-氧化锆质耐火材料 mullite-zirconia refractories

2.5.2 术语句子

用矾土和煤矸石合成**刚玉莫来石材料**。

Corundum-mullite material synthesized from bauxite and coal gangue.

传统的**莫来石陶瓷**广泛用于陶器、白瓷和瓷器，以及钢铁、有色金属、水泥玻璃生产和化学工业中的耐火材料。

Traditional **mullite ceramics** are widely used for pottery, whiteware, and porcelain, and for refractories in the steel, nonferrous metal, cement glass producing and in the chemical industry.

氧化锆增韧莫来石（ZTM）是优良的高温结构陶瓷材料。

Zirconia toughened mullite（ZTM）ceramics is one of finest performance high temperature structure ceramics.

特别是**莫来石-氧化锆复合材料**，由于具有良好的性能，如韧性、化学稳定性和高抗蠕变性，是重要的技术应用材料。

Particularly **mullite-zirconia composites** are materials with important technological applications due to their good properties such as toughness, chemical stability, and high-creep resistance.

作为氧化铝和铝硅酸盐质陶瓷的**莫来石**是传统耐火材料应用中最常用的材料。

Mullite as alumina and aluminiosilicate ceramics are among the most used materials for traditional refractory applications.

以天然矿物制备莫来石材料时，为了提高莫来石材料的一致性和可靠性，采用均化工艺技术制备了**矾土基均化莫来石熟料**。

To improve quality consistency and reliability of mullite materials fabricated by natural minerals, the technology of homogenized process was adopted to prepare **bauxite-based homogenized mullite grogs.**

大量工作通过反应烧结氧化铝-锆英石混合物的方法获得**莫来石-氧化锆复合材料**。

Much work has been carried out for obtaining **mullite-zirconia composites** by the reaction sintering of alumina-zircon mixtures.

莫来石纤维制品在高达约 1000 ℃的高温下表现出优异的绝缘特性，例如，用于管道或回转窑密封，以及用于金属熔炼炉中电极的电绝缘。

Mullite fiber fabrics display excellent high-temperature insulation characteristics up to about 1000 ℃, for example, for tube or rotary kiln seals, and for the electrical insulation of electrodes in metal smelting furnaces.

刚玉莫来石砖是一种应用广泛的耐火制品，主要用于高炉陶瓷杯垫、炉墙和风口，具有良好的抗铁水侵蚀、抗碱侵蚀以及抗热震性。

Corundum mullite brick is one of the widely used refractory products, mainly used in blast furnace ceramic cup pad, walls and tuyere, because of the good resistance to hot metal erosion, alkali corrosion as well as thermal shock.

在**刚玉-莫来石材料**中加入 ZrO_2 可以提高材料的热强度和抗蠕变性，并显著改善抗热震性。

ZrO_2 addition to **corundum-mullite materials** would lead to increase in hot strength and creep resistance as well as marked improved in thermal shock resistance.

传统的**莫来石陶瓷**广泛用于陶器、白色器皿和瓷器，以及钢铁、有色金属、水泥、玻璃生产和化学工业的耐火材料。

Traditional **mullite ceramics** are widely used for pottery, whiteware, and porcelain, and for refractories in the steel, nonferrous metal, cement, glass producing and in the chemical industry.

莫来石晶须对多孔莫来石陶瓷的孔隙结构有显著影响，增加了多孔莫来石陶瓷的开孔率和总孔隙率。

The pore structure of the porous mullite ceramics was significantly affected by the **mullite whiskers** which increased the open porosity and total porosity.

对于高杂质的**电熔融莫来石**，也可采用铝矾土或 Al_2O_3 与高岭石的混合物。较高的 Al_2O_3 含量可以通过快速淬火和较低的 Al_2O_3 整体成分来实现，或者通过非常缓慢的冷却过程来实现。

For **fused-mullite** of lower quality, bauxites or mixtures of Al_2O_3 and kaolinite have also been used. Higher Al_2O_3 contents can be achieved by rapid quenching and lower Al_2O_3 bulk composition or, alternatively, by a very slow cooling process.

单相莫来石陶瓷用于耐火材料、坩埚、热电偶管、热交换器、硅太阳能电池器件的衬底、牙科陶瓷元件、热气体过滤器、电子封装等。

Monolithic **mullite ceramics** are used in refractories, crucibles, thermocouple tubes, heat exchangers, substrates for silicon solar cell devices, dental ceramic components, hot gas filters, electronic packaging, etc.

莫来石涂层用于避免材料氧化，采用化学气相沉积（CVD）方法沉积一层莫来石薄涂层。莫来石涂层也可应用于油水清洗和热气体容器的基底。

Mullite coatings are used to avoid oxidation of materials, and a thin coating of mullite is deposited using the chemical vapor deposition (CVD) method. Mullite coatings are also used on substrates for oily water cleaning and hot gas.

2.5.3 术语段落

莫来石陶瓷的开发将为世界上大量存在的工业废粉煤灰带来巨大的经济价值，而目前这些工业废粉煤灰大多没有得到很好的利用。此外，廉价的天然原料铝土矿被证明是工业氧化铝生产莫来石的有效替代品。这一特征支持了直接使用天然原料的实用性，而不是通过复杂的工艺生产氧化铝[20]。

The development of **mullite ceramics** would add great economic value to industrial waste fly ash that existed abundantly throughout the world, most of which are not well utilized currently. Moreover, inexpensive raw natural mineral bauxite was testified as an effective substitute for industrial alumina for the production of mullite. This fact supports the practicability of the direct use of natural raw materials, instead of alumina produced by the complicated process[20].

具有适当比例（75∶25或25∶75）两种晶相的**刚玉-莫来石材料**比纯刚玉或纯莫来石材料具有更好的高温强度性能。在刚玉-莫来石材料中加入氧化锆可以提高材料的高温强度和抗蠕变性。在刚玉-莫来石材料中引入TiO_2，添加量在刚玉和莫来石中的固溶度范围内，会降低烧结温度，并在高温下保持良好的力学性能[21]。

Corundum-mullite materials with two crystalline phase in appropriate proportion (75∶25 or 25∶75) are better in high temperature strength properties than straight corundum or mullite. ZrO_2 addition to corundum-mullite materials would lead to increase in hot strength and creep resistance. TiO_2 addition to corundum-mullite materials within its solid solubility limit in corundum and mullite would lower sintering temperature and retain the good mechanical properties at high temperatures[21].

莫来石砖是应用最广泛的高炉炉缸内衬材料之一，具有诸多优点。研究了不同热处理温度下莫来石砖的显微结构和物理性能。结果表明，温度不影响莫来石骨料的微观形貌，但直接影响莫来石基质的显微结构。随着温度的升高，细晶粒的结合程度变高，基体中的结合体变得光滑。不同热处理后体积密度变化不大。

随着温度的升高，抗压强度和抗折强度增加，这有利于莫来石砖的抗渣/铁冲刷性能。不同热处理温度下莫来石砖的永久线数值均为负值，且相对较小，反映出莫来石砖体积稳定性的优异性能[22]。

Mullite brick is one of the most extensively utilized blast furnace hearth lining materials that feature several advantageous properties. In this study, the microstructure and physical properties of a mullite brick under different thermal treatment temperatures were investigated. The results showed that the temperature could not influence the micro-morphology of the mullite aggregate, but it directly affects the microstructure of the mullite matrix. With the increase in temperature, the degree of bonding of fine grains gets higher and the bonding body becomes smooth in the matrix. The bulk density changes little after different thermal treatments. With the increase in temperature, the compressive strength and modulus of rupture increase, which are beneficial for the mullite brick to the washing resistance of slag/iron. The permanent linear values of mullite brick are negative and relatively small under different thermal treatment temperatures, reflecting an excellent performance of the volume stability of mullite brick[22].

锆英石或电熔锆莫来石（骨料或粉末）加入矾土-碳化硅质耐火材料中，导致其导热系数降低。锆英石在 1500 ℃下不能分解，导致化学键弱，力学性能差。相比之下，含**电熔锆莫来石**的试样具有更好的常温耐压强度[23]。

Addition of zircon or fused zirconia-mullite (aggregate or powder) resulted in the decrease of thermal conductivity as compared to bauxite-SiC refractories. The presence of zircon is not able to decompose at 1500 ℃, leading to the weak chemical bond and inferior mechanical properties. By comparison, specimens containing **fused zirconia-mullite** possess a better cold crushing strength[23].

然而，合成**莫来石晶须**的长径比通常小于 20。此外，原料中出现的碱氧化物、氧化铁、氧化钠等氧化物以及添加剂（包括 V_2O_5、$NaH_2PO_4 \cdot 2H_2O$、Y_2O_3、La_2O_3、MnO_2 和 MoO_3）的引入会在莫来石晶须中留下一些残留的杂质元素，其中大部分对耐火材料的高温性能是不利的。为了在热冲击后获得更好的性能，有必要使用优化的原料和添加剂原位合成更大的莫来石晶须（长径比在 30 以上）[24]。

However, the aspect ratio of those as-synthesized **mullite whisker** is usually less than 20. Furthermore, those alkali oxide, iron oxide, sodium oxide and other oxides appeared in the raw materials and the introduction of additives (including V_2O_5、

$NaH_2PO_4 \cdot 2H_2O$、Y_2O_3、La_2O_3、MnO_2 and MoO_3) would leave some residual impurity elements in the as-prepared mullite whiskers, most of which are undesirable to the high temperature properties of refractory materials. For the sake of achieving better property after thermal shock, it is necessary to in situ synthesize larger mullite whiskers (with an aspect ratio of 30 or above) using optimized raw materials and additives without the problems above mentioned[24].

莫来石不仅对传统陶瓷具有重要意义，而且由于其良好的性能，已成为高级结构陶瓷和功能陶瓷的首选材料。莫来石的一些突出性能是低热膨胀、低导热、优异的抗蠕变性、高温强度高和良好的化学稳定性。莫来石的形成机理取决于含氧化铝和含硅的反应物结合的方法。它还与反应导致莫来石形成的温度（莫来石化温度）有关[25]。

Besides its importance for conventional ceramics, **mullite** has become a choice of material for advanced structural and functional ceramics due to its favorable properties. Some outstanding properties of mullite are low thermal expansion, low thermal conductivity, excellent creep resistance, high-temperature strength, and good chemical stability. The mechanism of mullite formation depends upon the method of combining the alumina-and silica-containing reactants. It is also related to the temperature at which the reaction leads to the formation of mullite (mullitisation temperature)[25].

在传统的加工方法中，莫来石粉末是成型后烧结得到的。这种莫来石被称为"**烧结莫来石**"。烧结莫来石一词描述了一种莫来石，它主要通过固态扩散控制反应从其原料中原位生成。氧化物、氢氧化物、盐和硅酸盐可用作原料。烧结温度、热处理时间、初始体成分、原料的性质、晶粒尺寸、混合效率以及 α-Al_2O_3 是否成核影响烧结石中 Al_2O_3 的含量。莫来化反应是通过铝、硅和氧原子的相互扩散，原料间发生固-固态反应或过渡液相态反应[25]。

In conventional processing methods, mullite powders are shape formed and sintered. This mullite is designated " **sinter-mullite** " . The term sinter mullite describes a mullite which has been produced from its starting materials essentially by solid-state diffusion-controlled reactions. Oxides, hydroxides, salts, and silicates can be used as the starting materials. The Al_2O_3 content of sinter mullites influenced by the sintering temperature, the duration of heat treatment, the initial bulk composition, by the nature, grain size and efficiency of mixing of the starting materials, and also whether α-Al_2O_3 nucleates. Mullitisation takes place by solid-solid or transient liquid-phase reactions of the starting materials by aluminium, silicon, and oxygen atom interdiffusion[25].

莫来化在起始物料的固相或液相反应（如黏土矿物、氧化铝和二氧化硅）中发生。纯氧化铝和纯二氧化硅是合成**莫来石陶瓷**最常用的原料。天然黏土矿物如瓷土或高岭土、叶蜡石、多晶硅线石、红柱石、蓝晶石等也可作为制造莫来石的原料[26]。

Mullitisation occurs during the starting materials′ solid or liquid phase reactions (e. g. , clay minerals, alumina, and silica). Pure alumina and pure silica are the most commonly used starting material for the synthesis of **mullite ceramics**. Clay minerals such as China clay or kaolin, pyrophyllite, polymorphs sillimanite, andalusite, kyanite, etc. , can be used as a starting material to fabricate mullite[26].

2.6 刚玉质耐火材料制品

2.6.1 术语词组

刚玉砖 corundum bricks
刚玉耐火材料 corundum refractories
板状刚玉 tabular corundum
再结合烧结刚玉砖 rebonded sintered corundum bricks
再结合电熔刚玉砖 rebonded electrically fused corundum bricks
熔铸刚玉砖 fused cast corundum bricks
熔铸铬刚玉砖 fused cast chrome-corundum bricks
熔铸锆刚玉砖 fused cast zirconia-corundum bricks
铬刚玉砖 chrome-corundum bricks
致密烧结刚玉陶瓷 dense sintered corundum ceramics
微孔刚玉砖 micropore corundum bricks
刚玉尖晶石质耐火材料 corundum-spinel refractories
锆刚玉耐火材料 corundum-zircon refractories
锆刚玉坩埚 corundum-zircon crucible
刚玉质透气砖 corundum purging plug
刚玉尖晶石透气砖 corundum-spinel purging plug
高炉用复合棕刚玉砖 composite brown corundum bricks for blast furnace

2.6.2 术语句子

以氧化铝-氧化铬为主要原料生产高性能**铬刚玉砖**。

High performance **chrome-corundum brick** was produced using synthetic alumina-chromia composite as main staring materials.

刚玉砖及**铬刚玉砖**的是石化行业造气炉和反应炉的关键耐火炉衬材料。

Corundum bricks and **chrome-corundum bricks** are the key refractories for gasifier and reaction furnace in petrochemical industry.

将 Cr_2O_3 添加到刚玉中可获得出色的高温强度以及更好的耐腐蚀性和耐磨性，使 **Al_2O_3-Cr_2O_3 耐火材料**在用作炉衬时能够抵抗熔渣和机械损坏。

The addition of Cr_2O_3 to corundum results in outstanding high-temperature strength as well as better corrosion and abrasion resistance, allowing **Al_2O_3-Cr_2O_3 refractory** to resist molten slag and mechanical damage when used as a furnace lining.

研制了树脂结合**铝碳砖**，并应用于中型高炉中段内衬。

Resin-bond **Al_2O_3-C brick** has been developed and applied for lining the middle-section of medium size blast furnaces.

电熔锆刚玉砖是玻璃熔窑重要的耐火材料。

Fused cast zirconia-corundum brick are the most commonly used refractory for glass tank furnaces.

高纯**刚玉砖**的特点是耐火度高、热强度高、体积稳定性好、高温化学惰性好。

The high purity **corundum bricks** are characterized by high refractoriness, hot strength, dimensional stability and chemical inertness at elevated temperature.

总结了精炼钢包用透气砖的损毁原因和现用**刚玉-尖晶石透气砖**在改善抗热震性方面的研究新进展。

Damage reasons of purging plug for refining ladle and research progress on thermal shock resistance of **corundum-spinel purging plug** were summarized.

传统**刚玉质耐火材料**有着优异的力学性能和耐火度，是高温熔炉衬里最广泛使用的材料之一。

Traditional **corundum refractories** are one of the most widely used materials for working linings of high-temperature furnaces due to their excellent mechanical properties and refractoriness.

因而，使用微孔刚玉骨料制备用于高温熔炉衬里的**轻量刚玉耐火材料**（**LCR**）。

Lightweight corundum refractories (**LCR**) for the working linings of high-temperature furnaces were therefore prepared using microporous corundum aggregates.

2.6.3 术语段落

碱性炉渣在耐火材料层表面的润湿特性直接影响耐火材料表面的包覆和腐蚀。实验结果表明，加热速率对渣在**刚玉耐火材料**表面的润湿过程有抑制作用，加热速率的增加提高了渣柱在耐火材料表面充分铺展所需的温度。改变加热气氛实验结果表明，在抑制渣柱在耐火材料表面的润湿性方面，还原气氛比氧化气氛更有效，这将渣完全润湿的温度从 870 ℃提高至 910 ℃。当耐火材料和水平面之间的角度增大到 15°~20°时，润湿性由于重力对黏附效果的影响而下降。实验结果表明，渣柱在高温下发生溶解并相互反应，形成包层、界面反应层和耐火基体层。结合热力学分析，生成高熔点的莫来石、霞石和长石的界面层，填充耐火材料表面的孔隙和缝隙，从而防止熔渣对耐火材料的侵蚀[27]。

The wetting characteristics of alkaline slag on the surface of the refractory layer directly affect the cladding and corrosion of the refractory surface. The experimental results show that the heating rate has an inhibitory effect on the wetting process of the slag column on the surface of **corundum refractories**, and the increase in heating rate leads to the increase in the temperature required for the slag column to fully spread on the surface of refractories. Changing the heating atmosphere shows that the reducing atmosphere is more effective than the oxidizing atmosphere in restraining the wettability of the slag column on the refractory surface, which increases the temperature of complete wetting from 870 ℃ to 910 ℃. When the angle between the refractory and horizontal planes increases to 15° and 20°, the wettability worsens due to the effect of gravity on the effect of adhesion. The experimental results show that the slag column dissolves refractorily at high temperature and reacts with each other, forming the cladding layer, interface reaction layer and refractory matrix layer. Combined with thermodynamic analysis, the interfacial layer of mullite, nepheline and feldspar with a high melting point is generated to stuff up the pores and slits on the surface of refractory materials, thus preventing the erosion of refractory materials by molten slag[27].

水泥结合刚玉耐火材料在使用过程中受到的热冲击缩短了其使用寿命。因此，改进耐火材料的抗热震性是决定材料质量的最关键因素。一些常见的方法，如水淬火方法（PRE/RS-1）、空气淬火方法（PRE/RS-2）等用于测量耐火材料的抗热震性。这些评估标准利用断裂强度的衰减，以表征材料的抗热震性，而不是取决于定量参数（例如，断裂能量）[28]。

Thermal shock during the service process of **cement-bonded corundum refractories** greatly shortens the service life. Hence, improving the thermal shock resistance of refractories is one of the most crucial factors that determine the material quality. Usually, there are a few popular methods, such as the water-quenching method (PRE/RS-1), the air quenching method (PRE/RS-2), etc., that are used to measure the thermal resistance of refractories. These evaluation criteria utilize the attenuation of rupture strength to characterize the thermal shock resistance of materials, rather than depend on the quantitative parameters (e. g., fracture energy) of fracture mechanics[28].

Al_2O_3-SiO_2-ZrO_2 系统中一类重要的耐火材料是以刚玉、莫来石和斜锆石为基础结构的。它广泛应用于前炉、进料器、玻璃熔窑，作为活塞、管道、通道、罩砖和导流环等装置中。这些材料被广泛用作耐火材料陶瓷生产用快速窑炉中的滚筒瓷砖、餐具和卫生用品。因此，**氧化铝-莫来石-氧化锆耐火材料**的许多工艺性能在很大程度上取决于相组成。例如，由于莫来石的热膨胀系数远低于刚玉，因此莫来石与刚玉相的比值的增加将提高抗热震性。另外，高温蠕变和耐碱性化合物腐蚀在很大程度上取决于玻璃相的量及其化学成分[29]。

An important category of refractories in the Al_2O_3-SiO_2-ZrO_2 system is based on corundum, mullite and baddeleyite structures. It is widely utilized in forehearth, feeders, glass melting furnaces, as plungers, tubes, channels, mantle blocks, and orifice rings. These materials are extensively used as refractory rollers in fast firing kilns for the manufacture of ceramic tiles, tableware and sanitaryware. Therefore, many technological properties of **alumina-mullite-zirconia refractories** greatly depend on phase composition. For example, it is expected that an increase of the mullite-to-corundum ratio will improve the thermal shock resistance, since mullite has a thermal expansion coefficient much lower than corundum. Moreover, creep at high temperature and resistance to corrosion by alkaline compounds depend to a large extent on the amount of glassy phase and its chemical composition[29].

刚玉-氧化锆复合材料作为一种有前景的结构材料一直被广泛关注。氧化铝的高弹性模量在四方 ZrO_2 的夹杂物处提供了所需的应力水平，并为多晶型转变设置了障碍。氧化物在高温下缺乏互溶性，这加速了材料的相增韧，并限制了 Al_2O_3 晶粒的生长。为了使转化增韧有效，需要 ZrO_2 分布的高度均匀性，这是通过组分（以盐溶液的形式）的剧烈湿混合，然后喷雾干燥或煅烧来实现的；或者可以使用溶胶-凝胶工艺[30]。

Corundum-zirconia composites as promising structural materials have been the subject of much research. The high elastic modulus of aluminum oxide provides the required stress level at inclusions of the tetragonal ZrO_2 and sets a barrier to polymorphic transformations. The oxides lack mutual solubility at high temperatures, which expedites the phase toughening of the material and sets a limit to the Al_2O_3 grain growth. For the transformation toughening to be effective, a high uniformity of ZrO_2 distribution is required, which is achieved through vigorous wet mixing of the components (in the form of a salt solution), followed by spray drying or calcination; alternatively, a sol-gel process can be used[30].

通过高温试验，研究了温度对高炉用**刚玉质耐火材料**显微结构和物理性能的影响。结果表明，高温下刚玉砖基质中原位反应生成莫来石和玻璃相。随着温度的升高，基质中逐渐形成网状结构，1500 ℃烧后基质中最终形成多孔整体。刚玉试样的常温耐压强度随着温度的升高先增大后减小。刚玉试样的抗弯强度在1150 ℃时最高，随着温度的进一步升高，刚玉试样的抗弯强度降低。刚玉样品的显气孔率在室温下最低，但温度超过 1150 ℃后，显气孔率在同一水平内波动[31]。

The influence of temperature on the microstructure and physical properties of **corundum refractory** for blast furnace are investigated by a high-temperature test. The results show mullite and glass phases are generated by the in situ reaction in the matrix of corundum brick under high temperature. With the increase in temperature, the net structure is gradually formed in the matrix, and the multi-porous entirety is finally formed in the matrix at 1500 ℃. The cold compressive strength of corundum sample increases first and then decreases with the increase in temperature. The flexural strength of corundum sample is the highest at 1150 ℃, while with further increasing of temperature, the flexural strength of corundum sample decreases. The apparent porosity of corundum sample is the lowest at room temperature, but the apparent porosity of corundum sample fluctuates within the same level after the temperature exceeded 1150 ℃[31].

为了提高**铬刚玉砖**的抗热震性，在铬刚玉砖配料中添加不同量的红柱石等量取代相同粒度的电熔白刚玉，经混练、成型、干燥、烧成后，检测其致密度、常温强度、常温耐磨性、高温抗折强度和抗热震性，并进行 XRD、SEM 和元素面扫描分析。结果表明：（1）红柱石莫来化反应的体积膨胀效应降低了铬刚玉砖的显气孔率；（2）由于莫来石的密度和硬度均低于刚玉的，砖中刚玉相比例的降低不仅导致体积密度下降，也导致常温强度和常温耐磨性能呈下降趋势；（3）基质中交错分布的柱状莫来石可有效改善试样的高温抗折强度和抗热震性；（4）综合考虑，红柱石添加量以不超过 18%为宜[32]。

In order to improve the thermal shock resistance of **chrome-corundum bricks**, different amounts of andalusite were added to the formulation of chrome-corundum bricks to replace the equivalent fused white corundum with the same particle size. After mixing, shaping, drying and firing, the density, cold strength, cold abrasion resistance, hot modulus of rupture and thermal shock resistance were tested, and the XRD, SEM and elemental surface scanning were carried out. The results show that: (1) the volume expansion effect of mullitization of andalusite reduces the apparent porosity of chrome-corundum bricks; (2) the density and hardness of mullite are lower than those of corundum, the decrease of corundum ratio in brick leads to the decrease of bulk density, the strength and the cold abrasion resistance; (3) the cross-distributed columnar mullite in the matrix can effectively improve the hot modulus of rupture and thermal shock resistance of the specimens; (4) considering comprehensively, the andalusite addition shall not exceed 18%[32].

为提高刚玉质弥散型透气砖的性能，以电熔白刚玉颗粒及细粉、α-Al_2O_3 微粉、Cr_2O_3 微粉等为原料，固定骨料与基质的质量比为 85：15，分别用质量分数为0、1%、2%、3%的 $MgCO_3$ 等量替代电熔白刚玉细粉制备了**刚玉质弥散型透气砖**，研究了 $MgCO_3$ 微粉加入量对其性能的影响。结果表明：加入 1%$MgCO_3$ 微粉时，试样常温耐压强度、常温抗折强度和高温抗折强度增加；继续增大加入量，试样的常温和高温强度下降，体积密度减小，显气孔率升高，且透气度变化不大。加入的 $MgCO_3$ 高温下分解生成的 MgO 与基质中的 Al_2O_3 反应生成尖晶石起增强作用，当加入量过多时，试样膨胀率和高温强度下降，这与体系中生成了低熔点相有关[33]。

To improve the properties of **corundum based dispersive purging plugs**, fused white corundum particles and fine powder, α-Al_2O_3 micropowder and Cr_2O_3 micropowder were used as raw materials, the mass ratio of aggregates to matrix was 85 : 15, and the fused white corundum fine powder was substituted with $MgCO_3$ by different amounts (0, 1%, 2% and 3%, by mass fraction). The effect of the $MgCO_3$ micropowder addition on the properties of the purging plugs was studied. The results show that with 1% $MgCO_3$ powder, the cold crushing strength, the cold modulus of rupture, and the hot modulus of rupture of the samples increase, while as the addition keeps increasing, the cold and hot strengths decrease, the bulk density decreases, the apparent porosity increases, and the air permeability slightly changes. MgO from the decomposition of $MgCO_3$ at high temperatures reacts with Al_2O_3 in the matrix to form spinel creating strengthening effect. With excessive $MgCO_3$, the expansion rate and hot strength of the samples decrease, which is related to the formation of low melting point phases in the system[33].

Al_2O_3-Cr_2O_3 耐火材料是连续固溶体，具有很好的耐腐蚀性和耐磨性。为了使其得到高效利用，本研究对不同刚玉来源的 Al_2O_3-Cr_2O_3 样品进行了显微结构和性能研究。刚玉的来源包括烧结板状刚玉、电熔白刚玉以及含少量 β-Al_2O_3 和 TiO_2 杂质的棕刚玉。力学测试结果表明，白刚玉的引入破坏了材料的物理结构，而棕刚玉的作用则相反。通过结合白刚玉和棕刚玉，Al_2O_3-Cr_2O_3 砖达到最佳结合强度，从而在固相烧结过程中通过表面扩散发生快速颈部生长[34]。

Al_2O_3-Cr_2O_3 refractories are completely substitutional solid solutions and exhibit better corrosion and abrasion resistance. To enable the comprehensive utilization of it, the microstructure and properties of Al_2O_3-Cr_2O_3 samples with different corundum sources were investigated in this study. The starting sources of corundum sources included sintered tabular corundum, fused white corundum, or brown corundum with minor impurities of β-Al_2O_3 and TiO_2. The results of mechanical test showed that the introduction of white corundum deteriorates the physical structure, while brown corundum acts in an opposite manner. The optimum bonding strength of the Al_2O_3-Cr_2O_3 brick was reached by combining white and brown corundum, whereby rapid neck growth occurred via surface diffusion during solid-phase sintering[34].

研究了加入 α-Al_2O_3 纳米粉对**高纯刚玉砖**高温强度和抗热震性的影响，即分别引入 0、0.5%、1%、2%和 3%的 α-Al_2O_3 纳米粉及 0、4%、8%和 12%的 α-Al_2O_3微粉的试样在 1300 ℃、1400 ℃、1500 ℃和 1600 ℃下保温 5 h 煅烧后测定高温抗折强度（1400 ℃）和抗热震性（$\Delta T = 1100$ ℃，水冷 1 次）。结果表明：同时加入 α-Al_2O_3纳米粉和 α-Al_2O_3微粉可以显著提高制品的高温强度，抗热震性也有一定改善；加入 1% α-Al_2O_3纳米粉和 8% α-Al_2O_3微粉，1500 ℃ 5 h 烧后试样的高温抗折强度（1400 ℃）达到 24.6 MPa[35]。

Effects of α-Al_2O_3 nano-powder additions on high temperature strength and thermal shock resistance of **high purity corundum brick** were investigated. Hot modulus of rupture at 1400 ℃ and residual strength ratio after thermal shock one cycle (ΔT = 1100 ℃, water cooling) were determined for corundum brick specimens with additions of 0, 0.5%, 1%, 2% and 3% α-Al_2O_3 nano-powder respectively and 0, 4%, 8% and 12% α-Al_2O_3 micro-powder after being fired at 1300 ℃, 1400 ℃, 1500 ℃ and 1600 ℃ for 5 h respectively. The results show that simultaneous additions of α-Al_2O_3 nano-and micro-powder would lead to noticeable increase in HMOR and some improvement in TSR. The HMOR of the specimen with 1% α-Al_2O_3 nano-powder and 8% α-Al_2O_3 micro-powder addition after being fired at 1500 ℃ was 24.6 MPa[35].

2.7 铝硅质耐火材料制品

2.7.1 术语词组

硅莫砖 guimo bricks
莫来石-堇青石棚板 mullite-cordierite decks
莫来石-堇青石窑具 mullite-cordierite kiln furniture
莫来石-堇青石质匣钵 mullite-cordierite saggar
莫来石-SiC-O′-Sialon 复合材料 mullite-SiC-O′-Sialon composites
水泥窑用硅莫红砖 GMH bricks for cement kilns
β-Sialon 结合刚玉砖 β-Sialon bonded corundum bricks
β-Sialon 结合刚玉碳化硅复合材料 β-Sialon bonded corundum-SiC composites

2.7.2 术语句子

莫来石-堇青石复合材料因其机械强度高、热震稳定性好、原料廉价、烧结温度低而得到广泛应用。

The **mullite-cordierite composite** has been widely used owing to its high mechanical strength, good thermal shock stability, inexpensive raw materials, and low sintering temperature.

以堇青石、莫来石、高岭土和氧化铝微粉为原料制备**堇青石-莫来石窑具材料**。

Cordierite-mullite kiln furniture materials were prepared by using cordierite, mullite, kaoline and aluminum oxide micro-powder as raw materials.

在**莫来石-堇青石匣钵**的生产过程中，通常在匣钵上加一层耐腐蚀保护层，以防止锂离子电池阴极材料与莫来石-堇青石基底接触。

During the production of **mullite-cordierite saggars**, manufacturers usually add a corrosion-resistant protective layer on the saggar to prevent contact between the Li-ion battery cathode material and mullite-cordierite substrate.

新形成的非氧化物相（SiC、Sialon）具有高熔点、高强度、优异的抗热震性和耐腐蚀性，弥补了矾土基莫来石中杂质的负面影响，因此**莫来石-SiC-O′-Sialon 复合材料**具有良好的高温性能，可用于水泥窑。

The newly formed nonoxides (SiC, Sialon) have high melting point, high strength, excellent TSR and corrosion resistance, which would compensate the negative effects of impurities in bauxite-based mullite, thus the **mullite-SiC-O′-Sialon composite** would have good high temperature properties for application in cement kilns.

鱼雷罐用 **β-Sialon 结合刚玉复合材料**的结构和性能。

Structure and properties of **β-Sialon bonded corundum composites** for torpedo car.

在 **β-Sialon 结合刚玉砖**中，主晶相刚玉颗粒构成了砖体的骨架结构，原位生成的 β-Sialon 晶体填充于刚玉间隙中，与粒状刚玉紧密结合，这种结构特征显著增强了复合材料的高温力学性能。

In **β-Sialon bonded corundum bricks**, crystals in-situ formed are well interlaced and interlocked with the main crystal phase corundum skeleton structure, resulting in noticeably strengthening the high temperature mechanical properties.

匣钵是承担锂离子电池正极材料烧结过程中的储运作用。大多数匣钵材料以**莫来石（$3Al_2O_3 \cdot 2SiO_2$）**和堇青石（$2Al_2O_3 \cdot 2MgO \cdot 5SiO_2$）为基础。

The saggar is the storage and transportation component in the sintering process of Li-ion battery cathode materials. Most saggar materials are based on **mullite ($3Al_2O_3 \cdot 2SiO_2$)** and cordierite ($2Al_2O_3 \cdot 2MgO \cdot 5SiO_2$).

莫来石-堇青石复合材料具有机械强度高、热震稳定性好、原料便宜、烧结温度低等优点，因而得到了广泛的应用。

The **mullite-cordierite composite** has been widely used owing to its high mechanical strength, good thermal shock stability, inexpensive raw materials, and low sintering temperature.

在生产**莫来石-堇青石匣钵**的过程中，通常在匣钵上添加耐腐蚀保护层，以防止锂离子电池正极材料与莫来石-青石基板之间的接触。

During the production of **mullite-cordierite saggars**, manufacturers usually add a corrosion-resistant protective layer on the saggar to prevent contact between the Li-ion battery cathode material and mullite-cordierite substrate.

第三种组分不会影响和污染锂离子电池正极材料的生产，并且需要与**莫来石-堇青石复合材料**形成紧密的结构。

The third component cannot affect and contaminate the production of Li-ion battery cathode materials and is required to form an intimate structure with the **mullite-cordierite composites**.

以金属硅、高铝矾土粉末和碳化硅晶粒等为原料，通过原位氮化反应烧结工艺，制备了矾土基 **β-Sialon-bond-SiC 复合材料**。

Through in situ nitridation-reaction sintering process, bauxite-based **β-Sialon-bonded SiC composites** are prepared by using metallic silicon, bauxite powders and silicon carbide grains, etc. as starting materials.

由于方石英和液相保护层的形成，**铝土矿基 β-Sialon 基 SiC 复合材料**具有良好的抗氧化性能。

Bauxite based β-Sialon-bonded SiC composites has good oxidation resistance owe to the formation of protective layer of cristobalite and liquid phase.

2.7.3 术语段落

基于当前节能、减排、环保的行业背景，为了提高锂离子电池正极材料烧结用匣钵的使用寿命，降低消耗，有必要对其危害进行分析。研究了**莫来石-堇青石匣钵**的作用机理，确定了一种有效的优化工艺。众所周知，实现材料所需性能的有效方法是配制具有复杂相组成的化合物。因此多组分材料是由不同组成相的复杂个体特性提供的[36]。

Based on the current industry background of energy conservation, emission reduction, and environmental protection, to improve the service life and reduce the consumption of the saggar for sintering cathode materials of Li-ion batteries, it is necessary to analyze the damage mechanism of the **mullite-cordierite saggar** and determine an effective optimization process. It is known that an effective method for achieving the required properties of materials is the formulation of compounds with complex phase compositions. Thus, the distinctive features of the multi-component material are provided by the complex individual properties of the different constituent phases[36].

基于莫来石和刚玉的耐火砖，通常用于铁矿石颗粒生产的回转窑，以及来自铁矿石颗粒制造窑的沉积物材料，在实验室规模的试验中用于研究耐火材料/沉积物反应以及沉积物组分对耐火砖的渗透行为。测试的材料既有单片形式，又有粉末形式。碱金属碳酸盐（含有钠和钾）被用作腐蚀剂，以提高反应动力。通过扫描电子显微镜表征了样品中难熔/沉积物界面的形态变化和活性化学反应。X 射线衍射表明，碱金属与砖中的莫来石发生反应，这在含钠的情况下比在含钾的情况下更明显。碱金属与耐火砖之间的反应形成了霞石（$Na_2O \cdot Al_2O_3 \cdot 2SiO_2$）、钾硅岩和亲钾岩（均为 $K_2O \cdot Al_2O_3 \cdot 2SiO_2$）和亮氨酸（$K_2O \cdot Al_2O_3 \cdot 4SiO_2$）等相。这些相的形成导致砖产生 20%~25% 的体积膨胀，从而加速其降解[37]。

Refractory bricks based on mullite and corundum, commonly used in rotary kilns for iron ore pellet production, and deposit material from an iron ore pellet production kiln, were used in laboratory scale tests to investigate refractory/deposit reactions and the infiltration of deposit components into the refractory bricks. The materials tested were in both monolithic form and in the form of powder. Alkali metal carbonates (containing sodium and potassium) were used as corrosive agents, to increase reaction kinetics. The morphological changes and active chemical reactions at the refractory/deposit interface in the samples were characterized by scanning electron microscopy. X-ray diffraction showed that alkali metals react with the mullite in the bricks, this being more pronounced in the case of sodium than potassium. Phases such as nepheline ($Na_2O \cdot Al_2O_3 \cdot 2SiO_2$), kalsilite and kaliophilite (both $K_2O \cdot Al_2O_3 \cdot 2SiO_2$), and leucite ($K_2O \cdot Al_2O_3 \cdot 4SiO_2$) were formed as a consequence of reactions between alkali metals and the refractory bricks. The formation of these phases causes volume expansions of between 20% and 25% in the brick materials, which accelerate degradation[37].

Sialon 材料是具有宽广性能的一类固溶体，Sialon 陶瓷在高温下具有良好的力学性能、抗热震性和抗氧化性，热膨胀系数小，化学稳定性高，耐腐蚀，在工程上获得广泛的应用。**Sialon 结合 SiC 耐火材料**，可使 SiC 耐火材料的强度、抗氧化性和耐磨性都有增强，并有良好的抗碱侵蚀性，可广泛用于炼铁高炉炉腹、炉腰和炉身下部等。Sialon 结合刚玉砖，是氧化物-非氧化物复合材料，可提高刚玉砖的力学性能、抗热震性以及抗铁水熔损性能，用于高炉炉缸陶瓷杯的性能也要大大优于刚玉莫来石和棕刚玉陶瓷杯。试验利用金属硅粉、金属铝粉、活性氧化铝氮化反应合成 β-Sialon，研究 Sialon 相不同目标 Z 值以及 β-Sialon 结合刚玉砖烧结剂[38]。

Sialon materials are a kind of solid solution with wide properties. Sialon ceramics have good mechanical properties, thermal shock resistance and oxidation resistance at high temperatures, small expansion coefficient, high chemical stability, corrosion resistance, and are widely used in engineering. **Sialon combined with SiC refractory materials** can enhance the strength, oxidation resistance, and wear resistance of SiC refractory materials, and have good alkali corrosion resistance. It can be widely used in the bosh, waist, and lower part of iron making blast furnaces. Sialon combined with corundum bricks is an oxide non oxide composite material that can improve the mechanical properties, thermal shock resistance, and molten iron loss resistance of corundum bricks. It is used for the performance of blast furnace hearth ceramic cups. It can also be greatly superior to corundum mullite and brown corundum ceramic cups. It is synthesized by nitridation reaction of metal silicon powder, metal aluminum powder and Activated alumina β-Sialon, study the different target Z values of Sialon phase and β-Sialon bonded corundum brick sintering agent[38].

以矾土基均质莫来石、硅粉和酚醛树脂为原料，制备了**莫来石-SiC-O′-Sialon 复合材料**。研究了不同 Si 添加量的样品在还原气氛下于 1500 ℃煅烧后的相组成、显微结构和性能。结果表明，莫来石-SiC-O′-Sialon 复合材料是在 1500 ℃碳包埋条件下，由 Si 和矾土基莫来石混合物经烧制而成。复合材料中的晶须状 SiC 和 O′-Sialon 是由 Si 与 C、CO 或 N_2 原位反应形成的，晶须的生长遵循 VS 和 VLS 机制。随着 Si 粉添加量的增加，复合材料的常温、高温强度和抗热震性都有所提高，最佳 Si 添加量（质量分数）为 10%。高温性能的提高是由于非氧化物晶须填充在莫来石骨架中，形成交叉连锁的网络结构，以及 SiC 晶须与莫来石直接结合，产生增强、增韧作用[39]。

Mullite-SiC-O′-Sialon composites were prepared using bauxite-based homogenized mullite, silicon powder and phenolic resin as raw materials. The phase composition, microstructure and properties of samples with various Si addition calcined at 1500 ℃ under reducing atmosphere were investigated. The results indicated that the mullite-SiC-O′-Sialon composites were obtained at 1500 ℃ in carbon embedded condition by firing Si and bauxite-based mullite mixtures. The whisker-like SiC and O′-Sialon in the composites were formed because of the reactions between Si and C, CO or N_2, and the growth of whiskers followed VS and VLS mechanisms. Both cold and high temperature strength, and thermal shock resistance of the composites increased with increase of silicon powder addition, and the optimized Si addition (mass fraction) was 10%. The improved properties at high temperature were due to the non-oxide whiskers filling in the skeleton of mullite and forming interlocking network structure as well as direct bonding between SiC whiskers and mullite, creating strengthening and toughening effects[39].

采用坩埚法对市售硅莫砖1680、硅莫砖1550、高铝砖、**低氧化铝莫来石砖**进行抗碱侵蚀对比试验，对侵蚀后坩埚进行物相及微观结构分析并探讨碱侵蚀行为。结果表明：（1）硅莫砖和低铝莫来石砖抗碱侵蚀性相对较好，抗剥落高铝砖抗碱侵蚀性能较差。（2）硅莫砖中的碳化硅在高温下氧化，在耐火砖表面形成致密层，可有效抑制K元素的侵蚀渗透，因此，材料的抗碱侵蚀性能优良；抗剥落高铝砖由于选用的原料品级较差，样砖显气孔率很高，碱渗透最深，并生成大量的钾霞石而导致试样产生大的碱裂，因此，材料的抗碱侵蚀性能很差。（3）低铝莫来石砖显气孔率低，所用原料中莫来石均质料及红柱石结构致密，且结构中存在一定数量的高硅非晶相，可与K元素反应生产高黏度玻璃相阻塞气孔，使得材料抗碱侵蚀性能较好[40]。

The alkali erosion resistance of commercially sold guimo bricks 1680, 1550, high alumina brick and **low alumina mullite brick** were tested by crucible method. The phase and microstructure of the bricks after erosion and the alkali erosion behavior were discussed. The results show that the alkali erosion resistance of guimo bricks and low alumina mullite bricks are relatively better, while the anti-peeling high alumina bricks are worse. The silicon carbide could be oxidized at high temperature forms a dense layer on the surface of the guimo brick, which can effectively inhibit the infiltration of K element, and the material has excellent alkali erosion resistance. The anti-stripping alumina brick, have high porosity and poor alkali-corrosion resistance due to poor raw materials, so a large amount of $KAlSiO_4$ formation volume expansion leads to cracking. The low alumina mullite brick has low porosity, and a certain amount of high silicon amorphous phase can react with K element to produce high viscosity glass phase to reduce porosity, inhibit the further infiltration of K element, so as to improve the alkali erosion resistance[40].

对**β-Sialon 结合刚玉砖**在0.5~5 MPa、0.8~8 MPa、1~10 MPa 三种循环载荷作用下，从常温到1500 ℃条件下的损伤行为进行了初步探索。结果表明：(1) 在应力幅为5 MPa下，从常温到800 ℃，β-Sialon 结合刚玉砖的弹性模量不随循环周次的增加而变化，属弹性变形范围；1000~1400 ℃弹性模量随循环周次增加逐渐增大，循环应力的作用不会加深材料的损伤；1500 ℃时弹性模量随循环周次增加逐渐减小并最终断裂。(2) 在1400 ℃，应力幅小于8 MPa时，β-Sialon 结合刚玉砖的弹性模量随循环周次增多而增大；应力幅提高到10 MPa时，出现弹性模量随循环周次增多而下降的性能弱化现象[41]。

The fatigue characteristics of **β-Sialon bonded corundum brick** under cyclic loading (0.5-5 MPa, 0.8-8 MPa, 1-10 MPa) from room temperature to 1500 ℃ was studied. The results show that, (1) under test conditions (the stress range is 5 MPa), with increasing loading cycle, no change in MOE was seen below 800℃ which belongs to the range of elastic deformation ; at 1000-1400 ℃, the MOE increases slightly, the effect of cyclic stress will not deepen material damage, and at 1500 ℃ the elastic modulus gradually decreases with increasing cycles and ultimately fractures. (2) At 1400 ℃ with stress amplitude < 8 MPa, the elastic modulus of β-Sialon bonded corundum bricks increases with the increase of cyclic cycles. When the stress range is up to 10 MPa, there is a performance weakening phenomenon where the elastic modulus decreases with increasing cycle times[41].

研究了**矾土基 β-Sialon 结合刚玉复合材料**在高温使用下的力学性能、抗热震性、抗氧化及抗 K_2CO_3 的侵蚀等性能，并和氧化铝基 β-Sialon 结合刚玉复合材料的性能进行了对比。结果表明：矾土基 β-Sialon 结合刚玉复合材料的高温抗折强度在相同条件下均高于氧化铝基 β-Sialon 结合刚玉复合材料；相同条件下，矾土基 β-Sialon 结合刚玉复合材料热震后的残余抗析强度均高于氧化铝基 β-Sialon 结合刚玉复合材料；β-Sialon 结合刚玉复合材料的氧化过程呈保护型氧化，温度高于1250 ℃时，矾土基 β-Sialon 结合刚玉复合材料的氧化速率比氧化铝基 β-Sialon 结合刚玉复合材料的低；在1300 ℃的 K_2CO_3 液中侵蚀6 h后，β-Sialon 结合刚玉复合材料的抗 K_2CO_3 侵蚀性能比刚玉砖和高铝砖好得多[42]。

Mechanical properties, thermal shock resistance, anti-oxidation and K_2CO_3-erosion resistance of **bauxite-based β-Sialon bonded corundum** were investigated and compared with those of alumina-based β-Sialon bonded corundum. Results show that the high temperature strength of rupture for bauxite-based β-Sialon bonded corundum is higher than that of alumina-based β-Sialon bonded corundum under the same conditions, and the bauxite-based β-Sialon bonded corundum shows a higher residue rupture strength after thermal shock than that of alumina-based β-Sialon bonded corundum in the same conditions. The oxidation process of β-Sialon/Al_2O_3 composite is a protective oxidation. The oxidation resistance of bauxite-based β-Sialon bonded corundum is better than that of alumina-based β-Sialon bonded corundum when the temperature is over 1250 ℃. Compared with high alumina brick and corundum brick, β-Sialon bonded corundum displays much better property of K_2CO_3-erosion resistance at 1300 ℃ for 6 h[42].

参考文献

[1] 邢东明，李勇，张秀华，等．热风炉高温区用硅砖中鳞石英的结构演变［J］．硅酸盐学报，2019，47（12）：1818-1824.

[2] Fan Jiahang, Li Yong, Gao Yuan, et al. Properties of both Chinese silica brick and silica raw material［J］. Ironmaking & Steelmaking, 2022, 49（5）：495-505.

[3] 张秀华，马晨红，钱雨，等．硅质耐火材料中矿化剂氮化硅铁和碳酸钙的作用机理［J］．硅酸盐学报，2023，51（3）：594-601.

[4] Fan Jiahang, Li Yong, Zhang Xiuhua, et al. Performance of silica bricks with ferrosilicon nitride as the mineralizer［J］. Ceramics International, 2022, 48：26791-26799.

[5] Pilate P, Lardot V, Cambier F, et al. Contribution to the understanding of the high temperature behavior and of the compressive creep behavior of silica refractory materials［J］. Journal of the European Ceramic Society, 2015, 35（2）：813-822.

[6] 徐平坤．腊石砖工艺特点及与相变的关系［J］．冶金丛刊，1997，3：28-55.

[7] 徐平坤．蜡石砖的显微结构与特性［J］．陶瓷，1993，1：14-18.

[8] Debnath N K, Pabbisetty V K, Sarkar K, et al. Preparation and characterization of semi-silica insulation refractory by utilizing lignite fly ash waste materials［J］. Construction and Building Materials, 2022, 345：128321.

[9] Li Kunming, Pan Chuancai, Xie Jinli, et al. Compressive creep behavior of low porosity fireclay bricks（LPFBs）for glass furnace regenerator［J］. Key Engineering Materials, 2016, 680：352-357.

[10] 周忠华．废耐火砖在烧结黏土砖中的有效利用［J］．砖瓦，2021，398（2）：15-18.

[11] 徐国涛，张彦文，杨帆，等．焦炉蓄热室黏土格子砖发泡变形的原因及机制分析［J］．耐火材料，2020，54（1）：61-65，69.

[12] 张永鹤，张岩岩．莫来石晶须的原位合成及对高铝黏土砖耐火材料的韧化效果［J］．轻金属，2020，499（5）：16-20.

[13] Palaniyappan S, Annamalai V E, Ashawinkumaran S, et al. Utilization of abrasive industry waste as a substitute material for the production of fireclay brick［J］. Journal of Building Engineering, 2022, 45：103606.

[14] 徐恩霞，钟香崇．高铝砖高温弯曲应力-应变关系［J］．耐火材料，2005，39：266-269.

[15] Wu Yang, Li Guangqiang, Tan Fangguan, et al. Research on creep damage model of high alumina bricks［J］. Ceramics International, 2022, 48（19）：27758-27764.

[16] Li Xiaohui, Gu Huazhi, Hang Ao, et al. Bonding mechanism and performance of rectorite/ball clay bonded unfired high alumina bricks [J]. Ceramics International, 2021, 47 (8): 10749-10763.

[17] Mcgee T D, Dodd C M. Mechanism of secondary expansion of high-alumina refractories containing calcined bauxite [J]. Journal of the American Ceramic Society, 1961, 44 (6): 277-283.

[18] 钟香崇. 产学研结合, 发展有中国特色的优质耐火材料 [J]. 耐火材料, 2009, 43 (1): 1-4.

[19] 钟香崇. 新一代矾土基耐火材料 [J]. 硅酸盐通报, 2006 (5): 92-98.

[20] 张艳利, 张小会, 程庆先, 等. 我国耐火原料的现状与发展 [J]. 耐火与石灰, 2023, 48 (1): 12-18.

[21] Zhong Xiangchong, Sun Gengchen. Thermomechanical properties of corundun-mullite-zirconia materials [J]. China's Refractories, 1988, 3: 3-10.

[22] Yao Shun, Zhou Heng, Wu Shengli, et al. Microstructure and physical properties of a mullite brick in blast furnace hearth: influence of temperature [J]. Ironmaking & Steelmaking, 2021, 48 (1): 55-61.

[23] Zhu Lingling, Li Sai, Gao Zexu, et al. Effect of in situ formed acicular mullite whiskers on thermal shock resistance of alumina-mullite refractories [J]. Journal of the Australian Ceramic Society, 2023, 59: 259-266.

[24] Hou Zhaoping, Liu Cheng, Liu Liangliang, et al. Microstructural evolution and densification behavior of porous kaolin-based mullite ceramic added with MoO_3 [J]. Ceramics International, 2018, 44 (15): 17914-17918.

[25] Anggono J. Mullite ceramics: its properties structure and synthesis [J]. Journal Teknik Mesin, 2005, 7 (1): 1-10.

[26] Roy R, Das D, Rout P K. A review of advanced mullite ceramics [J]. Engineered Science, 2021, 18: 20-30.

[27] Cheng Guishi, Zhao Tengfei, Zhao Ying, et al. Wetting, cladding and corrosion properties of alkaline slag on dense corundum refractories [J]. Ceramics International, 2022, 48 (4): 5795-5804.

[28] Pan Liping, He Zhu, Li Yawei, et al. Inverse simulation of fracture parameters for cement-bonded corundum refractories [J]. JOM, 2019, 71: 3996-4004.

[29] Zanelli C, Dondi M, Raimondo M, et al. Phase composition of alumina-mullite-zirconia refractory materials [J]. Journal of the European Ceramic Society, 2010, 30 (1): 29-35.

[30] Suvorov S A, Turkin I A, Dedovets M A. Microwave synthesis of corundum-zirconia materials [J]. Refractories and Industrial Ceramics, 2002, 43: 283-288.

[31] Yao Shun, Wu Shengli, Zhou Heng, et al. Influence of temperature on the microstructure and physical properties of corundum refractory brick in the blast furnace hearth [J]. Ironmaking & Steelmaking, 2020, 47: 263-270.

[32] 占化生, 李金雨, 李燕京, 等. 红柱石加入量对铬刚玉砖性能的影响 [J]. 2019, 53:

461-463.
[33] 邱鑫，张仕鸣，贾全利，等. $MgCO_3$ 微粉对刚玉质弥散型透气砖性能的影响 [J]. 耐火材料，2021，55 (3)：190-193.
[34] Pan Dunxiang, Zhao Huihong, Zhang Han, et al. Effect of different corundum sources on microstructure and properties of Al_2O_3-Cr_2O_3 refractories [J]. Ceramics International, 2019, 44 (15): 18215-18221.
[35] 贾晓林，钟香崇. α-Al_2O_3 纳米粉对高纯刚玉制品高温力学性能的影响 [J]. 2006, 40: 1-3, 6.
[36] Sun Ziheng, Yu Jun, Zhao Huizhong, et al. Damage mechanism and design optimization of mullite-cordierite saggar used as the sintering cathode material in Li-ion batteries [J]. Journal of the European Ceramic Society, 2022, 42 (13): 6255-6263.
[37] Stjernberg J, Olivas-Ogaz M A, Antti M L, et al. Laboratory scale study of the degradation of mullite/corundum refractories by reaction with alkali-doped deposit materials [J]. Ceramics International, 2013, 39 (1): 791-800.
[38] 张莎莎，曲殿利. 高炉陶瓷杯用 β-Sialon 结合刚玉砖的研制 [J]. 冶金能源，2011，30 (2)：55-58.
[39] An Jiancheng, Ge Tiezhu, Xu Enxia, et al. Preparation and properties of mullite-SiC-O-Sialon composites for application in cement kiln [J]. Ceramics International, 2020, 46 (10): 15456-15463.
[40] 马淑龙，康剑，杨晨，等. 水泥窑用不同铝硅系耐火砖抗碱侵蚀性能研究 [J]，耐火材料，2023，57 (2)：131-134.
[41] 徐恩霞，张恒，钟香崇. 高温循环载荷作用下 β-Sialon 结合刚玉砖的损伤行为 [J]. 耐火材料，2008，283 (4)：261-263.
[42] 张海军，刘战杰，钟香崇. β-Sialon 结合刚玉复合材料的性能 [J]. 硅酸盐学报，2005，11：10-15.

3 碱性耐火材料制品

3.1 方镁石质耐火材料制品

3.1.1 术语词组

电熔镁砂（电熔氧化镁）fused magnesia
轻烧镁砂（轻烧氧化镁）caustic magnesite（light-burned magnesia）
烧结镁砂（死烧镁砂）sintered（dead burned）magnesia
冶金镁砂 metallurgical magnesia
镁砖 magnesia brick
烧成镁砖 fired magnesia brick
再结合镁砖 rebonded magnesite brick
化学结合镁砖 chemical bonded magnesite brick
焦油结合镁砖 tar-bonded magnesite bricks

3.1.2 术语句子

为了满足市场需求和降低成本，**电熔镁砂**工业过程开始关注能耗高、污染严重、原料利用率低这些问题。

In order to meet the market requirements and reduce costs，**fused magnesia** industrial process begins to focus on these issues：high energy consumption，serious pollution，low utilization of raw materials.

在目前的应用和研究中，MgO 由**轻烧镁砂**或白云石提供，而 $MgCl_2$ 通常来自卤水。

In present applications and researches，the MgO was provided by **caustic magnesite** or dolomite while the $MgCl_2$ is often from brines.

该研究提出了 0~6 mm 的烧结颗粒，其主要尺寸对应于**烧结镁砂**的大多数耐火最终产品中使用的尺寸。

This study proposes sintering particles of 0-6 mm in primary sizes corresponding to those employed in most refractory end-products of **sintered magnesia**.

膨胀计研究显示，当**死烧镁砂**与工业锐钛矿而不是金红石结合时，其膨胀较小（1%与 7.2%），并且反应发生较早（965 ℃与 1120 ℃）。

Dilatometric studies revealed that when **dead burned magnesia** is combined with industrial anatase instead of rutile, it expanded less（1% versus 7.2%）and the reaction occurred earlier（965 ℃ compared to 1120 ℃）.

结果表明，在还原熔炼过程中，氧化铝和镁砖的溶解由直接溶解到熔渣中转变为间接溶解，**镁砖**的溶解量小于铝砖。

The results show that during the reduction smelting, the dissolution of alumina and magnesia bricks changed from direct dissolution into the molten slag to indirect dissolution, and the amount of **magnesia bricks** dissolved was less than that of aluminum bricks.

3.1.3 术语段落

例如，由于其耐高温和透光性，**电熔镁砂**在航空航天领域被用作耐高温光学材料。热稳定性使电熔镁砂成为炼钢过程中重要的炉衬材料。为了获得具有上述良好特性的电熔氧化镁，在制造过程中通常使用具有高温电弧的电熔氧化镁炉（FMF）。FMF 是冶炼电熔镁砂的主要设备。大多数 FMF 的电气类型是三相交流（AC）电弧炉，少数类型是直流（DC）电弧炉。FMF 利用电弧将电能转化为热能，原料（轻烧氧化镁粉或菱镁矿原矿）被高温熔化形成熔池。冶炼结束时，FMF 的熔池冷却结晶，最终形成电熔氧化镁[1]。

For example, due to its high temperature resistance and light transmittance, **fused magnesia** is used as a high temperature resistant optical material in the aerospace field. And the high temperature resistance and thermal stability make fused magnesia an important material of the furnace lining for the steelmaking process. In order to obtain the fused magnesia having the above-mentioned good characteristics, the fused magnesia furnace（FMF）having high temperature arcs is usually used in the manufacturing process. The FMF is the main equipment to smelt fused magnesia. Most of the electrical types of FMFs are the three-phase alternating current（AC）arc furnaces, and a small number of types are the direct current (DC) arc furnaces. FMF used electric arc to convert electric energy into heat energy and the raw materials (light burned magnesia powder or raw magnesite) are melted by the high temperature to form a molten pool. At the end of smelting, the molten pool in FMF is cooled down to crystallize and eventually forms fused magnesia[1].

以**轻烧镁砂**和白云石为原料，制备了不同活性氧化镁含量的氯镁石水泥(MOCs)。测定了流动性、抗压和抗折强度。之后，用 XRD 和 SEM 对 MOCs 进行了分析。结果表明，以轻烧镁砂和白云石为原料制备的 MOCs 具有良好的工程性能。MOCs 中活性 MgO 的最低含量（质量分数）应为 33.4%。研究表明 MOC 的耐水系数与水化和水解反应的平衡相关，这提供了两种主要的改性途径，即提高水化速率或防止逆反应。同时，MOCs 的抗压强度与耐水系数之间没有明显的联系[2]。

Magnesium oxychloride cements (MOCs) with different amount of active magnesium oxide were prepared with **caustic magnesite** and dolomite. The fluidity, compressive and flexural strength were measured. After that, the MOCs were analyzed by XRD and SEM. The results indicated that MOCs prepared with caustic magnesite and dolomite obtained a good engineering performance. The result suggested that the minimum active MgO content (mass fraction) used in MOCs should be 33.4%. At the same time, this study shows that the water resistance coefficient of MOC is concerned with the balance of hydration and hydrolysis reactions which provides two main modification approaches, improving the hydration rate or preventing the reverse reaction. Meanwhile, it is shown in the experiment that there is no significant relationship between the water resistance coefficient and the compressive strength of MOCs[2].

研究了**高温烧结镁砖**在单轴压缩条件下的初始蠕变阶段行为。进行了与使用相关的载荷和温度下的压缩蠕变试验。使用 MATLAB 中的 Levenberg-Marquardt 优化算法对实验研究中获得的蠕变变形进行逆分析。该迭代反演方法用于确定诺顿-贝利蠕变定律的三个参数 a、n 和 K。因此，使用了相同温度但不同负载的三条曲线。正在研究从 1150 ℃到 1500 ℃的温度范围。该材料在 1200 ℃和 8 MPa 载荷下开始显示蠕变应变。使用有限元分析（FEA）并考虑诺顿-贝利蠕变模型，数值再现了单轴试验。对特定温度下的实验和模拟总应变曲线进行了比较，显示出良好的一致性。在加热过程中，耐火结构的三维部件是进一步热力有限元分析的对象。目的是通过使用线性弹性材料衬里模型与诺顿-贝利蠕变模型比较钢壳中的冯米塞斯应力。假设加热过程中压缩应力占主导地位，不考虑拉伸蠕变行为。此外，蠕变模型中假设体积不变。FEA 的结果显示了钢壳中的冯米塞斯应力从大约 430 MPa 显著降低到 150 MPa，这对于钢壳的强度来说更加现实[3]。

The primary creep stage behavior of **a fired magnesia brick** at high temperatures under uniaxial compressive conditions was investigated. Compressive creep tests under service-related loads and temperatures were conducted. The obtained creep deformations of the experimental investigations were subjected to an inverse analysis using the Levenberg-Marquardt optimization algorithm in MATLAB. This iterative inverse method was used to determine the three parameters a, n and K for the Norton-Bailey creep law. Therefore, three curves at the same temperature but with different loads were used. The temperature range from 1150 ℃ up to 1500 ℃ was under research. The material started to show creep strain at 1200 ℃ under a load of 8 MPa. The uniaxial test was reproduced numerically using finite element analysis (FEA) and considering the Norton-Bailey creep model. The experimental and simulated total strain curves for a specific temperature were compared and showed good agreement. A three-dimensional part of a refractory structure was the subject of a further thermo-mechanically finite element analysis during the heat-up process. The objective was to compare the von Mises stresses in the steel shell by using a linear elastic material model for the lining compared with a Norton-Bailey creep model. Under the assumption that compressive stresses are predominant during heat-up, tensile creep behavior was not considered. Moreover, volume constancy was assumed in the creep model. The results of the FEA showed a significant reduction in the von Mises stresses in the steel shell from about 430 MPa to 150 MPa, which is much more realistic referring to the strength of the steel shell[3].

本研究提出了一种新的方法，通过微波烧结镍铁渣和**烧结镁砂**的混合物，并添加高达 10%（质量分数）的氧化铝，以提高优质耐火材料的制备。结果表明，在微波烧结过程中，适量添加氧化铝有助于形成颗粒细小且相对均匀的镁橄榄石。它还促进了高熔点镁铝尖晶石（$MgAl_2O_4$）、镁铁铝尖晶石（$MgFe_{0.6}Al_{1.4}O_4$）和镁铝铬酸尖晶石（$MgAl_{0.5}Cr_{1.5}O_4$）的产生，这些尖晶石取代了在不添加氧化铝的烧结过程中形成的镁铬尖晶石（$MgCr_2O_4$），最终改善了所得耐火材料的耐火性和其他性能。通过向矿渣和 25%（质量分数）烧结镁砂的混合物中加入 4%（质量分数）氧化铝，在 1250 ℃下烧结仅 20 min，所得耐火材料获得 1790 ℃的耐火度。与在没有氧化铝的情况下制备的材料相比，耐火度增加了 156 ℃[4]。

This study presents a new approach to enhance preparation of superior-quality refractory materials by microwave sintering of the mixture of ferronickel slag and

sintered magnesia with addition of alumina up to 10% (mass fraction). It was shown that in the process of microwave sintering, the proper addition of alumina could contribute to formation of forsterite with fine and relatively uniform particle size. It also promoted the generations of high melting point magnesium aluminate spinel ($MgAl_2O_4$), magnesium iron aluminate spinel ($MgFe_{0.6}Al_{1.4}O_4$), and magnesium aluminum chromate spinel ($MgAl_{0.5}Cr_{1.5}O_4$) which replaced magnesium chromate spinel ($MgCr_2O_4$) formed during sintering without addition of alumina, eventually improving refractoriness and other properties of the resulting refractory material. By adding 4% (mass fraction) alumina to the mixture of slag and 25% (mass fraction) sintered magnesia for sintering at 1250 ℃ in only 20 min, the resulting refractory material obtained refractoriness of 1790 ℃. Compared with that prepared in the absence of alumina, the refractoriness was increased by 156 ℃[4].

近年来，冶金行业对基础轻质隔热材料，尤其是低导热**镁质耐火材料**有着巨大的需求。因此，本文探讨了通过合成轻质骨料和不同的颗粒组成来制备**微孔镁质耐火制品**，以满足轻质碱性耐火材料的供应。对 1600 ℃烧结的微孔镁质耐火制品的显微结构、孔径分布、相组成、热膨胀系数、导热系数和烧结性能进行了表征。结果表明，菱镁矿细粉（致孔剂）热分解产生的原始盐假象为轻质骨料的合成提供了均匀的微孔结构。通过调节致孔剂的含量来控制方镁石相的微孔形态。微孔镁质耐火材料的平均孔径为 1.5 ~ 4.2 μm，显气孔率从 29.88%提高到 32.46%。同时，导热系数从 0.037 W/(m · K) 增加到 0.217 W/(m · K)，表明引入同系造孔剂可以生产出高气孔率、低导热系数的轻质碱性耐火材料[5]。

In recent years, there was a huge demand for basic lightweight insulation materials in the metallurgical industry, especially **magnesia-based refractories** with low thermal conductivity. Therefore, the preparation of **microporous magnesia-based refractory products** through the synthesis of lightweight aggregate and the different grain compositions to meet the supply of light basic refractories is discussed in this paper. The microstructure, pore size distribution, phase composition, coefficient of thermal expansion, thermal conductivity and sintering properties of the microporous magnesia-based refractory products sintered at 1600 ℃ were characterized. The results indicated that the original salt pseudomorph produced by the thermal decomposition of magnesite fine powder (porogenic agent) provides a uniform microporous structure for the synthesis of lightweight aggregates. The microporous morphology of the periclase phase was controlled by adjusting the content of the porogenic agent. The average pore size of microporous magnesia-based refractory ranged from 1.5 μm to 4.2 μm, and the apparent porosity increased from 29.88% to 32.46%. In the same time, the thermal conductivity increased from 0.037 W/(m·K) to 0.217 W/(m·K), indicating that the introduction of homologous porogenic agents could produce lightweight alkaline refractories with high porosity and low thermal conductivity[5].

本文首次揭示了铝热还原TBFS过程中耐火材料的溶解平衡及其对铝热还原的影响。结果表明，与碳砖相比，氧化铝砖和镁砖能更有效地获得块状钛硅铝合金，并避免钛硅铝合金的高质量损失。此外，氧化铝砖和**镁砖**的侵蚀随着加入CaO含量的增加而增加；然而，与镁砖相比，氧化铝砖的腐蚀更严重。当使用镁砖时，钛的提取率最大（最大值为99.85%）。研究结果表明，镁砖是铝热还原钛硅铝合金的最佳耐火材料。这项工作为铝热还原钛硅铝合金的工业应用提供了重要的实验信息[6]。

Herein, for the first time, the dissolution equilibrium of refractories during the aluminothermic reduction of TBFS and its effect on aluminothermic reduction were revealed. The results revealed that the alumina and magnesia bricks were more effective for obtaining bulk Ti-Si-Al alloy and avoiding high mass loss of the Ti-Si-Al alloy compared to the carbon bricks. Furthermore, the corrosion of alumina and **magnesia bricks** increased with an increase in the content of the added CaO; however, the corrosion of the alumina bricks was more severe compared to the magnesia bricks. In addition, the largest extraction ratio of Ti (maximum value: 99.85%) was achieved when magnesia bricks were employed. The results of this study indicate that magnesia bricks are the optimal refractory for the preparation of Ti-Si-Al alloy via the aluminothermic reduction of TBFS. This work provides important experimental information for the industrial application of the aluminothermic reduction of TBFS in the preparation of Ti-Si-Al alloys[6].

碱性耐火材料的高温性能是在非氧化气氛中测定的。将罩式炉改造为合金惰性气体喷吹炉。测定了1500 ℃以下耐火材料的负荷和压缩蠕变。测试了碳砖、石墨砖、转炉用**焦油结合镁砖**以及高炉出铁沟和陷阱孔混合物[7]。

High temperature properties of **basic refractories** were determined in a non-oxidising atmosphere. Cover-type furnaces were reconstructed to alloy inert gas injection. Refractories under load and creep in compression were determined up to 1500 ℃. Carbon bricks, graphite bricks, **tar-bonded magnesite bricks** for converters as well as blast furnace runner and trap hole mixtures were tested[7].

3.2 白云石质耐火材料制品

3.2.1 术语词组

白云石耐火材料 dolomite refractory; dolomitic refractory
白云石耐火砖 dolomite refractory brick
稳定性白云石耐火砖 stabilized dolomite refractory brick
半稳定白云石耐火砖 semi-stablized dolomite refractory brick
镁质白云石耐火材料 magnesite-dolomite refractory; magnesia-dolomite refractory

3.2.2 术语句子

研究表明，和未添加纳米尖晶石的**白云石耐火材料**试样相比含有尖晶石纳米颗粒的白云石耐火材料能够有效阻止水化反应并能够生成致密的微观结构。

It is shown that spinel nano-particles containing **dolomite refractories** effectively prevent the hydration and create a dense microstructure compared to dolomite refractory specimen without nano-sized spinel added.

由于在钢水纯净度、环保意识以及资源短缺等方面不断提高的需求，**镁质白云石耐火砖**已经成为有吸引力的制钢用耐火材料之一。

Magnesite-dolomite refractory bricks have become one of the attractive steel making refractories due to the increasing demands of molten steel purity, the awareness of environmental protection and resource shortage grows.

众所周知，在**镁质白云石耐火材料**系统中由于 Ca 和 O 之间的弱结合力游离石灰石和大气中的水分会快速反应。

As known, in the **magnesia-dolomite refractory** system, the free-lime reacts rapidly with moisture from the atmosphere due to the weak bond between Ca and O.

稳定性**白云石耐火材料**的生产基于通过引入石英砂或者镁硅酸盐等氧化硅成分将白云石中的游离 CaO 束缚成为硅酸二钙或者硅酸三钙等高度耐火的形式，同时稳定硅酸二钙。

The production of stabilized **dolomite refractories** is based on binding free CaO of the dolomite into highly refractory forms of dicalcium or tricalcium silicates by introducing a silica component, for example, quartz sand or a magnesia-silicate component, with a simultaneous stabilization of dicalcium silicate.

3.2.3 术语段落

通常情况下，**镁质白云石耐火砖**由 50%~80%（质量分数）的氧化镁组成。有两种方法来制备镁质白云石耐火砖。第一种方法利用菱镁矿（$MgCO_3$）和白云石（$Mg \cdot Ca(CO_3)_2$）熔融烧结熟料作为初始原料来生产镁质白云石耐火砖，通过这种途径可以得到性能良好的同质产品。另外一种方法是混合菱镁矿和白云石并在高温下进行煅烧来制备镁质白云石耐火砖。镁质白云石耐火砖有许多优点，包括高耐火度、良好的热震稳定性、高温下高抗化学侵蚀性、有碳条件下的热力学稳定性以及合适的耐磨性。镁质白云石耐火砖被广泛应用于不同的工业领域，包括冶金和水泥窑。并且，由于有利于从钢水中除杂，镁质白云石耐火砖是公认的能够加工洁净钢产品的有效耐火材料之一[8]。

Typically, **magnesite-dolomite refractory bricks** are composed of 50%-80% (mass fraction) of magnesia (MgO). There are two ways to produce magnesite-dolomite refractory bricks. The first way is using fused and sintered clinker of magnesite ($MgCO_3$) and dolomite ($Mg \cdot Ca(CO_3)_2$) as starting material to produce the magnesite-dolomite refractory bricks which would results in more homogenous products with more favorable properties. Another way is mixing magnesite and dolomite together and calcined them at high temperature that let to produce an in-situ magnesite-dolomite refractory bricks. These refractory bricks have a lot of advantages such as high refractoriness, good thermal shock resistance, high chemical resistance at high-temperature, thermodynamic stability in the presence of carbon, and an appropriate abrasion resistance. Magnesite-dolomite refractory bricks are widely used in different industries such as metallurgy and cement kilns. Also, magnesite-dolomite refractory bricks have been considered to be one of the effective refractory types for processing clean steel products, due to these refractory bricks are beneficial to removing inclusions from molten steels[8].

镁质白云石耐火材料是在冶金工业应用最为广泛的碱性耐火材料，并开始在其他工业领域包括石灰、玻璃和水泥崭露头角。根据文献记载，**白云石质耐火材**

料可能替代水泥工业用的镁尖晶石和镁铬砖。这些耐火材料被认为是对人体健康和环境有益的无铬耐火材料。白云石质耐火材料具有独特的性质，包括高耐火度（>2200 ℃）、可接受的硬度以及合适的机械强度。另外，白云石质耐火材料具有良好的耐磨和抗侵蚀性。由于其原料易获得和成本，它是一种经济型材料[9]。

Magnesia-dolomite refractories are considered the most used basic refractories in the metallurgical industries and have started to emerge in other industries such as lime, glass, and cement. According to the literature, **dolomitic refractories** might be substitutes for magnesia-spinel and magnesia-chromite bricks for the cement industry. These refractories have been considered chrome-free refractories with benefits to human health and the environment. Dolomitic refractories have unique properties such as high refractoriness (>2200 ℃), acceptable hardness, and suitable mechanical resistance. Besides, dolomitic refractories possess good abrasion and high corrosion resistance. It is an economical material due to the availability and cost of its raw material[9].

3.3 方钙石质耐火材料制品

3.3.1 术语词组

铝酸钙耐火材料 calcium aluminate refractories
铝酸钙水泥 calcium aluminate cement (CAC)
一铝酸钙 calcuim monoaluminate ($CaO \cdot Al_2O_3$, CA)
二铝酸钙 calcuim dialuminate ($CaO \cdot 2Al_2O_3$, CA_2)
六铝酸钙 calcium hexaluminate ($CaO \cdot 6Al_2O_3$, CA_6)
七铝酸十二钙 $12CaO \cdot 7Al_2O_3$ ($C_{12}A_7$)
富铝铝酸钙耐火材料 alumina-rich calcium aluminate refractory

3.3.2 术语句子

事实上，**铝酸钙耐火材料**的性能与产品质量密切相关；此外，产品的质量直接离不开制备工艺。

Actually, the properties of the **calcium aluminate refractories** are closely related to the quality of products; furthermore, the quality of products is directly inseparable from the preparation technology.

在加热过程中，脱水反应会导致生成气孔，从而降低弹性模量；尽管如此，高温下生成的**二铝酸钙**和**六铝酸钙**恰恰相反能够增加弹性模量。

During the heating process, the dehydration reactions will bring about the porosity formation, which will decrease their elastic moduli; however, the formation of **CA_2** and **CA_6** at higher temperature could inversely increase their elastic properties.

3.3.3 术语段落

在冶金工业领域，**铝酸钙（$CaO-Al_2O_3$）**长期被用作优异的除硫除磷精炼渣。由于其独特的物理和化学性质，它们可以发挥和 CaO 耐火材料类似的作用捕获硫、磷和其他在钢水中的杂质，从而达到净化钢水的目的。此外，铝酸钙水泥通常比 CaO 耐火材料拥有更高的熔点、低热膨胀系数，以及在还原气氛下优异的热震稳定性和抗侵蚀性能。这些表现直接表明铝酸钙材料有望成为 CaO 耐火材料的替代产品并在冶金过程中充分发挥其独特优势[10]。

Calcium aluminates（$CaO-Al_2O_3$） have long been used as an excellent refining slag for desulfurization and dephosphorization in the metallurgical industry. Due to their unique physical and chemical properties, they can play a similar role as the CaO refractory to capture [S], [P], and other impurities in molten steel, achieving the purpose of purifying molten steel. Furthermore, calcium aluminate refractories usually possess higher melting points than CaO refractories, low coefficient of thermal expansion, and excellent thermal shock resistance and corrosion resistance in reducing atmosphere. Such behaviors directly exhibit that the calcium aluminate material is expected to become an alternative product of CaO refractory and give full play to its unique advantages in the metallurgical process[10].

在高温条件下具有高强度和良好的热震稳定性，氧化铝质耐火浇注料被广泛应用于炼钢过程中。作为其中一种应用最为广泛的水硬性耐火结合剂，**铝酸钙水泥（CAC）**对于这些耐火浇注料的性质尤为重要。商用铝酸钙水泥的主要矿物成分为**一铝酸钙（$CaO \cdot Al_2O_3$，**$w=40\% \sim 70\%$）、**二铝酸钙（$CaO \cdot 2Al_2O_3$，**$w<25\%$）和**七铝酸十二钙（$12CaO \cdot 7Al_2O_3$，**$w<3\%$）。这些矿物相的水化过程开始于水和这些材料颗粒表面开始接触，并产生多种水化相，比如 C_3AH_6（$Ca_3Al_2O_6 \cdot 6H_2O$），AH_3（$Al_2O_3 \cdot 3H_2O$），还有 CAH_{10}（$CaAl_2O_4 \cdot 10H_2O$）和 C_2AH_8（$Ca_2Al_2O_5 \cdot 8H_2O$）亚稳定相。在加热过程中，这些水化产物将逐渐脱水并和氧化铝反应，在 1200 ℃生产二铝酸钙，在 1400 ℃以上生成**六铝酸钙**。六

铝酸钙的生产会大幅地增强材料的热震稳定性以及高温下的力学强度，这是由于裂纹偏转和桥接等增韧机制的发展。另外，其在高铁渣中的低溶解度对于浇注料的抗侵蚀能力也有改善作用[11]。

With the high strength and excellent thermal shock resistance at elevated temperature, alumina-based refractory castables are widely used in the steel-making process. **Calcium aluminate cement (CAC)**, as one of the most widely used refractory hydraulic binders, is of great importance to the properties of these castables. The main mineral phases of commercial CAC are **CA ($CaO \cdot Al_2O_3$**, w=40%-70%), **CA_2 ($CaO \cdot 2Al_2O_3$**, w<25%), and **$C_{12}A_7$ ($12CaO \cdot 7Al_2O_3$**, w<3%). The hydration process of these mineral phases starts when water comes into contact with their particle surface, resulting in the generation of various hydration phases such as C_3AH_6 ($Ca_3Al_2O_6 \cdot 6H_2O$), AH_3 ($Al_2O_3 \cdot 3H_2O$), and some metastable compounds of CAH_{10} ($CaAl_2O_4 \cdot 10H_2O$) and C_2AH_8 ($Ca_2Al_2O_5 \cdot 8H_2O$). During the heating process, these hydration products will dehydrate gradually and react with Al_2O_3 to form CA_2 at 1200 ℃ and **CA_6** above 1400 ℃. The formation of CA_6 will significantly improve the thermal shock and creep resistance and the mechanical strength at high temperature due to the development of toughening mechanisms such as crack deflection and bridging. Furthermore, its relative lower solubility in high iron-containing slag should benefit the improvement on corrosion resistance for the castables[11].

3.4 尖晶石质耐火材料制品

3.4.1 术语词组

镁铝尖晶石砖 magnesia-spinel bricks

镁铁铝尖晶石砖 magnesia hercynite bricks

方镁石-镁铝尖晶石砖 periclase-magnesia alumina spinel bricks

镁铝碳砖 magnesia-alumina-carbon (MAC) bricks

镁铝钛砖 magnesia-alumina-titania bricks

氧化镁-镁铝尖晶石耐火材料 magnesia-magnesium aluminate spinel refractory

镁铝铬复合尖晶石砖 magnesia alumina chrome composite spinel bricks

方镁石-镁铝尖晶石耐火砖 periclase-magnesium aluminate spinel refractory (PMAS)

刚玉镁铝尖晶石 corundum-magnesia alumina spinel

3.4.2 术语句子

方镁石的线膨胀系数大，为 13.5×10^{-6} ℃$^{-1}$，弹性模量高，热膨胀大，对应力的缓冲能力差，而**镁铝尖晶石**的线膨胀系数（7.6×10^{-6} ℃$^{-1}$）低于氧化镁和氧化铝。

The periclase has a large expansion coefficient of 13.5×10^{-6} ℃$^{-1}$, high modulus of elasticity, high thermal expansion, and poor cushion ability to stress, while the expansion coefficient of **magnesia alumina spinel** (7.6×10^{-6} ℃$^{-1}$) is lower than that of magnesia and alumina.

众所周知，在水泥回转窑中，白云石砖比**镁尖晶石砖**具有更好的挂窑皮能力，这与不同涂层形成机制有关。

It is well known that doloma bricks present better coating adherence than **magnesia-spinel bricks** when applied in cement rotary kilns, which is related to the different coating formation mechanism.

为了承受这些苛刻的条件，**方镁石-镁铝尖晶石耐火材料**通常用于上部过渡区，因为它们具有优异的热机械性能（强度和抗热震性）和通常良好的耐腐蚀性。

To withstand these harsh conditions, **periclase-magnesium aluminate spinel refractories** are usually used in the upper transition zone, because they have excellent thermomechanical properties (strength and thermal shock resistance) and generally a good corrosion resistance.

设计了具有两种典型 Al_2O_3 骨料的**水泥结合刚玉尖晶石（Al_2O_3-$MgAl_2O_4$）预制耐火砖**作为耐火衬里。

A cement-bonded corundum-spinel (Al_2O_3-$MgAl_2O_4$) pre-cast refractory brick with two typical Al_2O_3 aggregates was designed as the refractory lining.

3.4.3 术语段落

方镁石耐钢水腐蚀，对含 CaO、FeO 的碱性渣有极好的抵抗力。特别是随着焙烧温度的升高和保温时间的延长，方镁石的晶粒尺寸增大。抗水化性和抗渣蚀性也有所提高。但方镁石的线膨胀系数大，为 13.5×10^{-6} ℃$^{-1}$，弹性模量高，热膨胀大，对应力的缓冲能力差，而**镁铝尖晶石**的线膨胀系数（7.6×10^{-6} ℃$^{-1}$）低于氧化镁和氧化铝。耐钢水腐蚀，抗热震性好。因此，选用大晶方镁石为主晶相，镁铝尖晶石为基体相[12]。

Periclase is resistant to molten steel and has excellent resistance to alkaline slag containing CaO and FeO. In particular, the grain size of periclase increases with the rising calcination temperature and the prolongation of the holding time. The hydration resistance and slag corrosion resistance are also improved. However, the periclase has a large expansion coefficient of 13.5×10^{-6} ℃$^{-1}$, high modulus of elasticity, high thermal expansion, and poor cushion ability to stress, while the expansion coefficient of **magnesia alumina spinel** (7.6×10^{-6} ℃$^{-1}$) is lower than that of magnesia and alumina. It is resistant to molten steel and has good thermal shock resistance. Therefore, large crystalline periclase was selected as the main crystalline phase and magnesia alumina spinel as the matrix phase[12].

可以确认，对于选定的熟料成分在运行期间砖表面涂层的形成和黏附取决于熟料与砖之间的物理和化学相互作用。当耐火材料的渗透性提高并且呈现良好的烧结度时，物理相互作用得到促进，这主要是由于使用了高烧成温度。这些特性证明对改善可涂覆性非常重要，因为高渗透性促进熟料渗透，而合适的烧结度促进耐火颗粒与熟料相的接触。对于化学相互作用，已经证明液相的形成对于黏附涂层是必不可少的。这样，添加2%（质量分数）的石灰石是将典型**镁尖晶石砖**的涂层强度从0 MPa提高到3 MPa最有效的方法。然而，由于$C_{12}A_7$和Q相的存在，这种添加导致1200 ℃和1485 ℃的热断裂模量下降。因此，结合良好黏附性和热性能的替代组合物将优选那些具有较高烧制温度并保持高水平耐火材料纯度的组合物[13]。

During the work, it was possible to confirm, with the selected clinker composition, that the coating formation and adherence on the brick surface depend on physical and chemical interaction between the clinker and the brick. The physical interaction was facilitated when the permeability of the refractory was elevated and when it presented good sintering degree, mainly due to the use of high firing temperature. These characteristics demonstrated great importance to improve coatability, since high permeability promotes clinker infiltration, whereas suitable sintering degree promotes the contact of refractory particles with the clinker phases. For chemical interaction, it has been demonstrated that the formation of liquid phase is essential to adhere coating. In this way, the addition of 2% (mass fraction) of limestone was the most efficient to improve the coating strength from 0 MPa to 3 MPa of a typical **magnesia-spinel brick**.

However, this addition resulted in a drop in hot modulus of rupture at 1200 ℃ and 1485 ℃ because of the presence of $C_{12}A_7$ and Q phase. Thus, alternative compositions which combine good adherence and hot properties would be preferably those with higher firing temperature and maintaining a high level of purity of the refractory[13].

选用6块砖在1500 ℃下进行50 h的转渣试验。根据耐火材料的最大腐蚀深度，对红土镍矿的耐蚀性依次为：铬刚玉砖、**镁铝铬复合尖晶石砖**、铬刚玉浇注料、**尖晶石铬刚玉砖**、镁铬砖、**镁铝尖晶石砖**。铬赋予铬刚玉砖和浇注料良好的耐蚀性，粘渣少。但在试验过程中，出现了厚度方向的裂纹。如果采用铬刚玉作为直接还原红土镍矿回转窑的炉衬材料，检修炉温控制将是一个很大的挑战。镁铝尖晶石砖具有良好的抗渗透性，但耐蚀性有限。镁铝铬复合尖晶石砖具有镁铝尖晶石砖和铬刚玉砖的优点，是最好的。富MgO尖晶石能吸收渗入的三氧化二铁，形成致密的铁镁尖晶石层，阻止熔融红土镍矿的渗入。因此镁铝铬复合尖晶石砖是红土镍矿直接还原回转窑的理想内衬[14]。

Six pieces of bricks are chosen for 50 h of rotary slag test at 1500 ℃. According to the maximum corrosion depth of refractories, the corrosion resistance against laterite nickel ore gets worse in the following sequence: chrome corundum brick, **magnesia alumina chrome composite spinel brick**, chrome corundum castable, **spinel chrome corundum brick**, magnesia chrome brick, and **magnesium aluminate spinel brick**. Chrome endows the chrome corundum brick and castable with good corrosion resistance and little slag adhesion. But during the test, cracks occur in the thickness direction. If the chrome corundum is adopted as the lining material of rotary kilns for direct reduction of laterite nickel ores, the control on the furnace cooling for overhaul will be a big challenge. The magnesium aluminate spinel brick has good penetration resistance but the limited corrosion resistance. The magnesia alumina chrome composite spinel brick is the best by possessing the advantages of magnesium aluminate spinel bricks and chrome corundum bricks. MgO-rich spinel can absorb the penetrated ferric oxide, and form a dense zeylanite layer, which prevents the penetration of the molten laterite nickel ores. So the magnesia alumina chrome composite spinel brick is an ideal lining of rotary kilns for direct reduction of laterite nickel ores[14].

腐蚀微观结构表明，在使用过的水泥结合**Al_2O_3-$MgAl_2O_4$耐火浇注料**上观察到一个深度约6900 μm的侵蚀区域，包括多孔反应层、渗透层和原始层。水泥结合Al_2O_3-$MgAl_2O_4$耐火浇注料经过工业试验后，在基体中观察到大量针状CA_6。相信CA_6形成于在服役期间耐火内衬上高温度梯度下的铝酸钙水泥和细氧化铝颗粒之间。在烧结和熔融的 Al_2O_3 聚集体上均观察到连续的内部 CA_6 层。然而，外层的形态各不相同。在烧结的 Al_2O_3 骨料上观察到约 70 μm 厚的 CA_2 和热铝尖晶石的外部混合层[15]。

Corroded microstructure revealed that a corroded region with a ~6900 μm depth, including a porous reaction layer, penetration layer and original layer, was observed on the used cement bonded **Al_2O_3-$MgAl_2O_4$ refractory castables**. A large number of acicular CA_6 was observed in the matrix of the cement-bonded Al_2O_3-$MgAl_2O_4$ refractory castables after industrial trials. It was believed that CA_6 formed between calcium aluminate cement and fine alumina grains under a high temperature gradient on refractory lining during service. A continuous inner CA_6 layer was observed both on the sintered and fused Al_2O_3 aggregates. However, the morphologies of outer layers varied. An outer mixed layer of CA_2 and hercynite with a ~70 μm thickness was observed on the sintered Al_2O_3 aggregate[15].

3.5 镁橄榄石质耐火材料制品

3.5.1 术语词组

镁橄榄石砖 magnesia forsterite bricks
镁橄榄石耐火砖 forsterite refractory brick
尖晶石-镁橄榄石 spinel-forsterite
纳米镁橄榄石/纳米镁铝尖晶石粉 nano forsterite/nano magnesium aluminate spinel powders
轻质镁橄榄石砖 light forsterite brick
方镁石-镁橄榄石轻质隔热耐火材料 periclase-forsterite lightweight heat-insulating refractories
镁硅砖 magnesia-silica brick; high-silica magnesite brick

3.5.2 术语句子

在泰国，有滑石和菱镁矿的自然资源。这两种原料主要是氧化镁和二氧化硅，因此具有合成**镁橄榄石耐火材料**和生产耐火砖的可能性。

In Thailand, there are natural sources of talc and magnesite. These two raw materials are mainly magnesia and silica, so there is interest and the possibility of synthesizing **forsterite refractory materials** and producing refractory bricks.

经过 40 h 的机械活化和随后在 1200 ℃下退火 1 h 后，获得了晶粒尺寸在 30~87 nm 范围内的纯**尖晶石-镁橄榄石纳米复合材料**。

Pure **spinel-forsterite nanocomposites** with crystallites size in the range of 30-87 nm were obtained after 40 h of mechanical activation and subsequent annealing at 1200 ℃ for 1 h.

镁橄榄石（forsterite）广泛用于制造业，尤其在耐火材料和铸造行业中发挥着重要作用。

Mg-olivine (forsterite) is used in the manufacturing industry vastly and plays a significant role especially in refractory and foundry industry.

改性 **EV/镁橄榄石复合材料**的导热系数估算与实验结果吻合较好，据估算，当改性 EV 的替换量（质量分数）为 100%时，改性 EV/镁橄榄石复合材料的导热系数为 0.157 W/(m·K)（1073 K）。

The estimated thermal conductivities of modified **EV/forsterite composite materials** show good agreement with that of experiments, and the thermal conductivity of modified EV/forsterite composite materials was 0.157 W/(m·K) (at 1073 K) in case the substitution rate of modified EV was 100% (mass fraction) through estimation.

3.5.3 术语段落

通过使用镁橄榄石作为镁砖的弹性成分，可以实现机械和热机械优化的预期目标。**镁橄榄石耐火材料**与窑原料和窑内气氛中的碱性化合物之间的稳定热化学相互作用是优势。降低熟料熔体渗透程度，优化包覆性，提高耐碱性。抗水合作用也增加了。自 2008 年以来，**镁橄榄石砖**在全球范围内与镁尖晶石砖相结合，成为众多水泥回转窑分段衬砌的利器[16]。

By the use of forsterite as elastifying component for magnesia bricks the intended target of mechanical and thermomechanical optimization is achieved. The stable thermochemical interaction of **magnesia forsterite refractory materials** with the kiln feed and alkali compounds from the kiln gas atmosphere is advantageous. The extent of clinker melt infiltration is reduced, the coatability is optimised and the alkali resistance is improved. Also the resistance to hydration is increased. Since 2008, **magnesia forsterite bricks** are used as an advantageous tool for sectional linings in numerous cement rotary kilns in combination with magnesia spinel bricks worldwide[16].

镁橄榄石耐火材料在泰国可以以滑石和菱镁石为原料合成。它们以 1∶5 的摩尔比混合并在 5 h 内机械活化，然后在 1300 ℃下煅烧 1 h。镁橄榄石晶体呈圆形，粒径小于 1 μm。在 1400 ℃下烧结 2 h 的镁橄榄石耐火砖（FB-14）具有最佳性能：零表观孔隙率，表观密度为 3.26 g/cm³，冷压强度为 72.18 MPa，导热系数（在 1000 ℃时）约为 2.1800 W/(m·K)，并且它们能够承受多达 39 个循环的热冲击。1400 ℃的烧结温度是制备耐火砖孔隙率为零的最低温度，因为 1300 ℃的烧结温度仍然出现轻微多孔。在耐蚀性方面，FB-14 耐火砖具有对铜渣和铅硅酸盐熔块的耐侵蚀能力。最重要的是，这些砖可以抵抗 18 次以上的熔融 SHMP 侵蚀循环[17]。

Forsterite refractory materials can be synthesized from talc and magnesite as raw materials in Thailand. They were mixed at a molar ratio of 1∶5 and mechanically activated at 5 h, then calcined at 1300 ℃ for 1 h. The forsterite crystals were round with less than 1 μm particle size. The forsterite refractory bricks sintered at 1400 ℃ for 2 h (FB-14) were the best properties: zero apparent porosity, an apparent density of 3.26 g/cm³, a cold crushing strength of 72.18 MPa, thermal conductivity (at 1000 ℃) was approximately 2.1800 W/(m·K), and they were able to withstand the thermal shock of up to 39 cycles. The sintering temperature of 1400 ℃ was the lowest temperature at which the refractory brick had zero porosity because the sintering temperature of 1300 ℃ still appears slightly porous. In terms of corrosion resistance, the FB-14 refractory bricks were corrosion-resistant to copper slag and lead silicate frit. Most importantly, the bricks were resistant to more than 18 cycles of corrosion by molten SHMP. In terms of corrosion resistance, the FB-14 refractory bricks were corrosion-resistant to copper slag and lead silicate frit. Most importantly, the bricks were resistant to more than 18 cycles of corrosion by molten SHMP[17].

基于 85%镁橄榄石和 15%尖晶石纳米粉末的烧结（1550 ℃）陶瓷体在烧结和力学性能方面表现出显著改善。CCS 达到最大值 333.78 MPa，比镁橄榄石块体试样高约 1 倍。当纳米尖晶石超过 15%时，由于表观孔隙率增加和堆积密度降低，CCS 降低。体积电阻率随着尖晶石含量增加至 15%而增加。陶瓷体中存在的低熔点相会减少晶粒间的接触面积，使电流更容易流动，并起到电流阻滞剂的作用。制备的**镁橄榄石/尖晶石陶瓷体**的突出特性使其在化学、工程和电子应用中非常有用[18]。

Sintered (1550 ℃) ceramic bodies based on nano powders of 85% forsterite and 15% spinel exhibited a pronounced improvement in sintering and mechanical properties. CCS reached the maximum value of 333.78 MPa, about one times higher than forsterite bulk specimens. Beyond 15% of nano spinel, CCS was decreased as a result of an increase in the apparent porosity and a decrease in the bulk density. The volume resistivity increases with increasing of spinel content up to 15%. The low melting phase exists in the ceramic bodies will reduce the inter grain contact areas for easier electrical current flow and acts as electrical current blockers. The outstanding features of the prepared **forsterite/spinel ceramic bodies** make them very useful in chemical, engineering and electronic applications[18].

使用滑石粉、氧化铝和碳酸镁粉末制备纯**尖晶石-镁橄榄石纳米复合材料**。除了机械活化外，仔细调整初始粉末的比例可防止形成不利于材料高温性能的顽火辉石相。制备的尖晶石-镁橄榄石纳米复合材料的晶粒尺寸范围为 30~87 nm。那些含有 10%和 20%尖晶石的样品的冷压碎强度高于那些没有任何尖晶石相的样品。随着镁橄榄石结构中 10%尖晶石的形成，CCS 达到最大值 67 MPa，比镁橄榄石大块样品高约 2 倍。由于裂纹的形成、表观孔隙率的增加、堆积密度的降低和部分尖晶石化反应，含有 20%~40%尖晶石样品的 CCS 发生了降低[19]。

Pure **spinel-forsterite nanocomposites** were prepared using talc, alumina, and magnesium carbonate powders. Carefully adjusting the ratio of initial powders besides mechanical activation prevented the formation of enstatite phase which is detrimental to the high temperature properties of material. The crystallites size of prepared spinel-forsterite nanocomposites was in the range of 30-87 nm. Cold crushing strength of those samples containing 10% and 20% spinel was higher than those samples without any spinel phase. With the formation of 10% spinel in the forsterite structure, CCS reached the maximum value of 67 MPa, about 2 times higher than forsterite bulk specimens. CCS decreased in those samples containing 20%-40% spinel as a result of the formation of cracks, an increase in the apparent porosity, a decrease in the bulk density, and partial spinellization reaction[19].

以**天然镁橄榄石**和菱镁石粉为起始原料，根据 M_2S 的化学成分，采用湿磨、半干法成型和不同温度煅烧制备样品。当煅烧温度超过 1300 ℃时，所有矿物相都会转化为所需的相。随着转速和焙烧温度的升高，样品的体积密度升高，显气孔率降低，冷压强度提高。综合考虑，400 r/min 和 1450 ℃为最优方案。无烟煤的加入使样品更轻，获得了一系列具有不同性能和均匀分布的微孔的轻质原料。利用所得轻质原料制备轻质耐火浇注料，具有良好的隔热性能[20]。

Samples were prepared using **natural forsterite** and magnesite powder as the starting materials according to chemical component of M_2S, wet milling, semi-dry molding and calcining at different temperatures. When the calcination temperature exceeds 1300 ℃, all the mineral phases convert to desired phases. With the increase of the rotation speed and the calcination temperature, the bulk density of the samples rises, the apparent porosity decreases and the cold compressive strength improves. By comprehensive consideration, 400 r/min and 1450 ℃ are taken as the optimal scheme. The anthracite addition makes the samples lighter, obtaining series of light-weight raw materials with different properties as well as uniformly distributed micro-pores. Light-weight refractory castables were prepared using the obtained light-weight raw materials, achieving good heating insulation[20].

膨胀蛭石改性后，复合材料的力学性能和隔热性能均得到显著提高。当改性膨胀蛭石的替代率为50%时，弯曲强度和压缩强度分别为11. 55 MPa 和22. 80 MPa，与未改性样品相比分别提高了 23. 8%和 44. 9%；导热系数为 0. 169 W/(m · K)(测试温度为 1073 K)，提高了 30. 5%。改性膨胀蛭石表面的活性多孔氧化铝气凝胶反应形成 $AlPO_4$，分布在膨胀蛭石的结构间隙中，从而获得良好的力学性能。这个过程在阻碍传热方面起到了非常重要的作用。**改性膨胀蛭石/镁橄榄石复合材料**的导热系数估算值与实验结果吻合较好，改性膨胀蛭石/镁橄榄石复合材料（改性膨胀蛭石的替代量为100%）在1073 K时的导热系数为0. 157 W/(m · K)，通过估算比未修改的样本提高了 39. 7%。如果有朝一日出现先进的生产工艺，这种改性膨胀蛭石含量高的复合材料可能是首选产品[21]。

After modification of expanded vermiculite (EV), both the mechanical and thermal insulation properties of composite materials were significantly improved. When the substitution rate of modified EV was 50%, the flexural and compressive strength were 11. 55 MPa and 22. 80 MPa, improved by 23. 8% and 44. 9%, respectively, compared with the unmodified sample; the thermal conductivity was 0. 169 W/(m · K) (at the testing temperature of 1073 K), improved by 30. 5%. The active and porous alumina aerogel on the surface of modified EV reacted to form $AlPO_4$, leading to good mechanical properties while being distributed in the structural gaps of the EV. This process played a very important role in obstructing heat transfer. The estimated thermal conductivities of modified EV/forsterite composite materials show good agreement with that of experiments, and the thermal conductivity of **modified EV/forsterite composite materials** (the substitution of modified EV was 100%) at 1073 K was 0. 157 W/(m · K), improved by 39. 7% compared with the unmodified sample through estimation. This kind of composite materials with high content of modified EV could be the product of choice if advanced production processing appears some day[21].

3.6 镁铬耐火材料制品

3.6.1 术语词组

镁铬砖 magnesia chrome brick; magnesia-chromite brick

直接结合镁铬砖 direct-bonded magnesia chrome brick

再结合镁铬砖 re-bonded magnesia chrome brick

半再结合镁铬砖 semi-rebonded magnesia chrome brick

化学结合镁铬砖 chemical-bonded magnesia chrome brick; unfired magnesia chrome brick

熔融镁铬砖 fused magnesia chrome brick; fused grain magnesia chrome brick

再生电熔粒状镁铬耐火砖 reconstituted fused-grain magnesia-chrome refractory brick

镁铬耐火材料 magnesia-chrome refractory

3.6.2 术语句子

Malfliet 等综述了在铜生产中耐火内衬的降解机理和应用，发现辉石矿渣引起的热化学负荷是**镁铬砖**常见的磨损机理。

Malfliet et al reviewed the degradation mechanisms and use of refractory linings in copper production, and found that the fayalite slag induced thermochemical load was a common wear mechanism of **magnesia-chromite bricks**.

Chen 等研究了**镁铬耐火材料**在铜熔炼炉中的稳定性，并发现高铬和高铁尖晶石的生成能够增加耐火内衬的稳定性。

Chen et al studied the stability of **magnesia-chrome refractories** in a copper smelting furnace and found that the formation of high-chrome and high-iron spinels could increase the stability of the refractory lining.

从一个二级铜冶炼厂收集了使用过的**直接结合镁铬耐火砖**以表征其在使用过程中发生的劣化。

Used **direct-bonded magnesia-chromite refractory bricks** from a secondary Cu smelter were collected to characterize the degradation occurring during application.

RH 精炼炉内衬的低位和高位部位主要采用熔融颗粒**再结合镁铬砖**和**直接结**

合镁铬砖。

The low vessel and upper vessel of the inner lining in RH refining furnaces mainly adopt fused grain **re-bonded magnesia-chromite bricks** and **direct-bonded magnesia-chromite bricks**.

半结合镁铬砖是一种镁铬部分熔融的耐火制品。

Semi-rebonded magnesia-chrome brick is a kind of refractory product with the partial fused magnesia chrome.

未烧结镁铬砖，也称为**化学结合镁铬砖**，其原料和制备过程与烧结镁铬砖基本相同。

Unfired magnesia chrome brick, also called **chemical-bonded magnesia chrome brick**, has basically the same raw materials and preparation process as fired magnesia chrome brick.

这项研究是一个更大的项目的一部分，目标是确定、预测和理解炼铜炉中**镁铬耐火砖**的劣化模式。

This study is part of a larger project in which the goal is to identify, predict and understand the degradation pattern of **magnesia-chromite refractory bricks** in copper-making furnaces.

3.6.3 术语段落

镁铬砖通常被用作铜和铅工业的冶炼厂的内衬，因为其能够抵抗火法冶金生产过程中的热负荷、机械负荷和化学负荷。事实上，铅铜锍的火法生产过程通常是一个还原过程，其中有大量的焦炭和铁填充物作为还原剂，这意味着镁铬砖是在更强的还原性气氛中服役[22]。

Magnesia-chrome bricks are commonly used as the lining of smelters in copper and lead industry because of their resistance against the thermal, mechanical and chemical loads in the pyrometallurgical production process. In fact, the pyrometallurgical production process of lead-copper matte is generally a reducing process with plenty of coke and iron fillings as reducing agent, which implies that the magnesia-chrome bricks service in stronger reducing atmosphere[22].

当从铜转炉炉渣中回收有价值的金属时，**镁铬耐火材料**作为电炉的内衬。然而，耐火材料的使用寿命不超过6个月。为了解决渗漏问题，通过采用静态腐蚀试验，研究了不同温度下浸泡在铜转炉渣中的镁铬耐火材料的腐蚀行为。首先对炉渣与耐火材料的溶解度和反应进行了热力学模拟。实验室腐蚀的样品表明，铁和硅渗透到耐火材料中，导致材料溶解和反应性腐蚀。同时，在炉渣-耐火材料界面上形成了一个富铁层，可以防止进一步腐蚀。鉴于 Fe_3O_4 的含量，认为1300 ℃是促进保护层形成的最佳温度。实验室的研究结果与模拟结果非常吻合。基于腐蚀行为，本研究提出了一种新的可以延长炉子的使用寿命方法[23]。

Magnesia-chrome refractories act as a lining for electric furnaces when valuable metals are recovered from copper converter slags. However, the lifetime of refractories does not exceed six months. To address the leakage problem, the corrosion behavior of a magnesia-chrome refractory immersed in a copper converter slag at different temperatures was investigated by adopting static corrosion test. The thermodynamic simulations of solubility and reactions of the slag with the refractory were first conducted. The laboratory corroded samples showed that Fe and Si penetrated the refractory material, resulting in material dissolution and reactive corrosions. Meanwhile, an Fe-rich layer formed at the slag-refractory interface which could prevent further corrosion. Given the Fe_3O_4 content, 1300 ℃ was concluded as the optimal temperature to facilitate the formation of the protective layer. The laboratory findings were in agreement with the simulation well. On the basis of the corrosion behavior, this study proposed a novel approach that prolongs the lifetime of furnaces[23].

在典型的炼铜温度（1300 ℃）下，用静态手指试验测试了六种镁铬矿和六种无铬耐火砖对铜和阳极炉渣的渗透性和耐腐蚀性。通过电子探针微分析（EPMA）和扫描电子显微镜（SEM）技术，研究了收到的和经过测试的耐火材料的微观结构。结果表明，除了由熔融刚玉和镁铝尖晶石组成的氧化铝基砖，以及由烧结氧化镁和锆石添加物制成的氧化镁基砖之外，样品的整体磨损率非常低。在所有类型的耐火材料中，由于炉渣和耐火材料的相互作用，形成了新的相。除了从最新一代的**直接结合镁铬砖**的铜区回收的样品外，其余的都被铜和渣成分（氧化铜、氧化铁、氧化铝和二氧化硅）完全渗透。然而，无铬型砖的渗透液量比镁铬砖高。对这种不同的渗透行为进行了解释。结果表明，对于阳极炉衬来说，经济上可行的无铬耐火材料替代品仍然难以找到[24]。

The penetration and corrosion resistance to copper and anode slag of six magnesia-chromite and six chrome-free refractory brick types were tested using static finger tests at a typical copper-refining temperature (1300 ℃). The microstructures of the as-delivered and tested refractory types were investigated by means of electron-probe micro-analysis (EPMA) and scanning electron microscopy (SEM) techniques. The results showed that the overall wear rate of the fingers was very low, with the exception of the alumina-based brick made of fused corundum and magnesia-alumina spinel, and the magnesia-based brick made of sintered magnesia and zircon addition. In all refractory types new phases were formed as a result of slag-refractory interactions. Apart from the samples recovered from the copper zone of the latest generation of **direct-bonded magnesia-chromite bricks**, all the rest were completely infiltrated by copper and slag components (copper oxide, iron oxide, alumina and silica). However, the amount of infiltrated liquid in the chrome-free types was higher than in the magnesia-chromite bricks. Explanations are provided for the distinct infiltration behaviour. The results show that economically viable chrome-free refractory alternatives are still elusive for anode furnace linings[24].

传统上，RH 脱气机的内衬是**再结合镁铬砖**，而**直接结合镁铬砖**和**半再结合镁铬砖**在低磨损区域更适宜。一般来说，铬镁耐火材料被用于钢铁冶金领域，因为它们具有良好的热稳定性和高温性能。通常，镁铬耐火内衬的使用寿命可以由其服役至失效期间钢铁处理的次数来决定。在 RH 浸渍管和 RH 装置的下部，存在最困难的温度和化学、机械和物理工作条件，会发生密集的钢流。因此，主要的破坏因素是：流动钢水的侵蚀，铁氧化物对耐火材料的腐蚀，与硅酸钙相渗透有关的耐火材料的腐蚀过程，相对较高的温度大大增加了对耐火材料内衬的腐蚀和侵蚀，定期真空脱气和真空还原条件导致的快速温度变化。一般情况下，耐火材料 RH 浸渍管内衬的使用寿命在 50~200 次钢处理之间[25]。

Traditionally, snorkels of RH degasser are lined with **re-bonded magnesia-chrome bricks** while **directly bonded magnesia-chrome bricks** and **semi-rebonded magnesia-chrome bricks** are preferred in lower wear regions. Generally magnesia-chrome refractories are used the sector of ferrous metallurgy because of their good thermal stability and high temperature performance. Generally, the lifetime of a magchrome refractory lining can be defined by the number of steel treatments to failure. Both, in the RH-snorkels and in the lower of parts of the RH installations, having the most difficult temperature and chemical, mechanical, and physical conditions of work, the intensive flow of steel occurs. Hence, the major destructive factors are: erosion by the molten flowing steel, corrosion of refractories by iron oxides, corrosion process of refractories connected with the infiltration of the calcium silicate phases, the relatively high temperature which significantly increases corrosion and erosion of refractory lining, the rapid temperature changes result from the periodic vacuum degassing and vacuum reducing conditions. Generally, the refractory RH-snorkel linings lifetime ranges between 50 and 200 steel treatments[25].

南非铬矿的杂质含量低，氧化铁含量高。中国需要进口相当数量的南非铬矿作为耐火原料。基于中国的菱镁矿资源，研究人员尝试用轻烧的菱镁矿和铬矿在超高立窑（约2000 ℃）中烧结制造镁铬熟料，发现合成的材料会黏附在窑内衬上。用轻烧镁砂和铬矿在超高竖窑中烧结的镁铬熟料的体积密度很低。因此，欧洲的全合成镁铬砖技术并不适合中国。中国已经独立开发和生产了高质量的熔融镁铬熟料，逐步形成了熔融半再结合和**熔融再结合镁铬砖**的体系[26]。

South African chrome ores have low impurity contents and high iron oxide contents. China needs to import a considerable amount of South African chrome ores as refractory raw materials. Based on the magnesite resources in China, the researchers tried to fabricate magnesia-chrome clinker with light burned magnesia and chrome ore sintering in ultra-high shaft kilns (at about 2000 ℃) and found that the synthetic material adhered to the kiln lining. The magnesia-chrome clinker sintered with caustic magnesia and chrome ore in ultra-high shaft kilns had low bulk density. So the European technology of fully synthetic magnesia-chrome bricks was not suitable for China. China had independently developed and produced high-quality fused magnesia-chrome clinker, gradually forming systems of fused semi-rebonded and **fused rebonded magnesia-chrome bricks**[26].

已经发表了一些关于**镁铬耐火材料**磨损机制的研究。Mosser等人描述了镁铬砖在Ruhrstahl-Heraeus（RH）脱气机中的劣化机制。他们观察到，磨损的主要特征是富含硅酸盐或铝酸盐的炉渣渗入，渗入物与砖的基体发生反应，以及随后渗入区被热侵蚀。致密区域倾向于在过渡区向非渗透区剥离[27]。

A number of studies of the wear mechanisms of **magnesia-chromite refractories** have been published. Mosser et al have described the degradation mechanisms of magnesia-chromite bricks in Ruhrstahl-Heraeus (RH) degassers. They observed that the wear is primarily characterized by the infiltration of silicate rich or aluminate containing slags into pores, reaction of the infiltrate with the matrix of the brick, and a subsequent hot erosion of the infiltrated zone. The densified zone tends to peel off in the transition zone to the non-infiltrated area[27].

3.7 镁锆质耐火材料制品

3.7.1 术语词组

镁锆砖 magnesia zirconia brick
镁锆耐火材料 magnesia-zirconia refractory
再结合镁锆砖 re-bonded magnesia zirconia brick
预合成氧化镁-氧化锆 presynthesized magnesia-zirconia
氧化镁-氧化锆陶瓷 $MgO-ZrO_2$ ceramic
氧化镁氧化锆复合材料 magnesia zirconia composite
氧化镁-氧化锆共混物 magnesia-zirconia co-clinker
镁基耐火材料 magnesia-based refractory
尖晶石镁锆砖 magnesia-spinel-zirconia brick

3.7.2 术语句子

在试验中，**再结合镁锆耐火材料**是由熔融镁和熔融共晶镁锆粉通过一定的技术手段制成的，并研究了烧结温度对镁锆砖性能的影响。

In the test, **re-bonded magnesium-zirconium refractory materials** is made of fused magnesia and fused eutectic magnesium-zirconium powder by certain techniques, and effect of sintering temperature on performance of magnesia zirconia brick is researched.

将烧制的**镁锆砖**、尖晶石镁砖、**尖晶石镁锆砖**和尖晶石镁钛砖砖与镁铬砖进行了比较，镁锆砖对高碱度矿渣表现出最好的抗侵蚀性。

Burnt **magnesia-zirconia bricks**, magnesia-spinel bricks, **magnesia-spinel-zirconia bricks** and magnesia-spinel-titania bricks were compared with magnesia-chrome bricks. Magnesia-zirconia bricks showed the best corrosion resistance to high basicity slag.

测试的材料是取自同一预分解水泥窑的镁尖晶石（MSp）和**镁锆（MZ）耐火材料**，热回转窑料（富含硫和氯）和波特兰熟料（富含硫）。

The tested materials were the magnesia-spinel (MSp) and the **magnesia-zirconia (MZ) refractories**, the hot kiln meal (rich in sulphur and chlorine) and the Portland clinker (rich in sulphur) taken from the same precalciner cement kiln.

Guo 等人讨论了 $ZrSiO_4$ 对烧结氧化镁的烧结以及由**预合成氧化镁氧化锆**组成的镁砖性能的影响。

Guo et al discussed the effect of $ZrSiO_4$ on sintering of sintered-magnesia and the properties of magnesia brick made of **presynthesized magnesia-zirconia**.

为了解决基于 ZrO_2 的惰性基质燃料的低导热性和基于 MgO 的惰性基质燃料在水中的不稳定性，建议将 **MgO-ZrO_2 双相陶瓷**作为轻水反应堆燃料的基质，用于锕系嬗变和钚的燃烧。

To address the low thermal conductivity of the ZrO_2-based inert matrix fuel and the instability in water of the MgO based inert matrix fuel, the **dual-phase MgO-ZrO_2 ceramics** are proposed as a matrix for light water reactor fuel for actinide transmutation and Pu burning.

3.7.3 术语段落

通过对比**镁锆砖**和镁铬砖的抗渣行为，并通过简化模型研究了镁锆砖的制备参数与抗 RH 渣腐蚀之间的关系。通过正交实验设计（OED）分析了影响镁锆砖抗渣性能的制备参数，如颗粒混合比、凝集剂、烧结温度、氧化锆含量和氧化锆品种。结果表明，烧结温度和氧化锆含量是两个关键的影响因素。通过最陡峭的上升法和中心成分设计（CCD）进一步考察了烧结温度和氧化锆含量对浸润深度比例的影响，并建立了镁锆砖被 RH 渣侵蚀的电阻腐蚀的统计分析模型[28]。

The relation between preparing parameters of **magnesia-zirconia bricks** and their resistance to RH slag corrosion was studied by contrasting the slag resistance behaviors of magnesia-zirconia bricks and magnesia-chromite bricks and by a simplified model. The preparing parameters that influence the RH slag resistance properties of magnesia-zirconia bricks, such as grain mixture ratio, agglutinate, sintering temperature, zirconia content, and variety of zirconia, were analyzed by orthogonal experimental design (OED). The results indicate that sintering temperature and zirconia content are two key influencing factors. The effects of sintering temperature and zirconia content on the ratio of infiltrate depth were further examined by the steepest ascent method and central composition design (CCD), and a statistic analysis model of resistance corrosion of magnesia-zirconia bricks eroded by RH slag was built[28].

20 世纪 80 年代，水泥回转窑烧结带和过渡带主要使用镁铬砖，因其性能优良，但其中的 Cr^{6+} 具有致癌性，对空气、水污染、动物或人类造成很大危害。因此，新的不含铬的基本耐火材料开始被开发。一些发达国家（如欧洲、美国和日本等）已停止在水泥回转窑烧结带使用镁铬砖，并严格限制或制定法律禁止使用镁铬砖，这加速了无铬基础材料的发展，他们开发了镁尖晶石砖、镁白云石砖、白云石砖等，但这些砖在水泥窑中的使用效果并不理想。本试验以熔融镁砂和熔融共晶镁锆粉为原料，通过一定的技术手段制成**再结合镁锆耐火材料**，并研究烧结温度对**镁锆砖**性能的影响[29]。

Nineteen eighties, the sintering zone of cement rotary kiln and transition belt use main magnesia chrome brick because of its excellent performance, but the Cr^{6+} is carcinogenicity, which causes great harm to the air, water pollution, animal or human. Therefore, new basic refractory material of free of chrome began to be developed. Some developed countries (for example, Europe, the United States and Japan etc.) have stopped to use magnesium chrome bricks at sintering zone in cement rotary kiln, and strictly limited or enacted the law to ban the use of magnesia chrome brick, which accelerates the development of basic materials free of chrome, and they have developed magnesium spinel brick, magnesia dolomite brick, dolomite brick etc, but the using effect of these bricks in cement kiln is not ideal. In the test, **re-bonded magnesium-zirconium refractory materials** is made of fused magnesia and fused eutectic magnesium-zirconium powder by certain technics, and research effect of sintering temperature on performance of **magnesia zirconia brick**[29].

嵌入氧化镁中的单斜锆石加强了**镁锆砖**的结构，通过氧化锆的相变产生微裂缝并吸收断裂能量以提高其抗热震性。但必须有至少 15%（质量分数）的单斜锆石加入氧化镁中以达到最佳功能。虽然较小量的氧化锆不能优化镁锆砖，但大约 3%（质量分数）的含 ZrO_2 镁锆砖在 RH 脱气装置的使用性能各有优劣[30]。

The monoclinic zirconia embedded in magnesia strengthens the structure of **magnesia-zirconia brick**, generating the microcracks through phase transformation of zirconia to absorb fracture energy and to improve its thermal shock resistance. But at least 15% (mass fraction) monoclinic zirconia has to be incorporated into magnesia, displaying optimal function. Although smaller amounts of zirconia could not optimize magnesia-zirconia bricks to be appropriate, about 3% (mass fraction) ZrO_2-containing magnesia-zirconia bricks experienced good or bad uses in RH degasser[30].

在本研究中，从烧制的废旧镁碳耐火材料中获得的二次氧化镁熟料制成的氧化镁和**氧化镁-氧化锆耐火陶瓷**与传统方式获得的材料（商业原料）进行了比较。研究发现，废旧镁碳耐火材料可以被成功回收利用，但值得注意的是，所获得的二次镁质熟料的质量较低。通过添加氧化锆和/或生产熔融氧化镁-氧化锆共混物可以减少开放性气孔并增加体积密度从而大大改善其质量。此外，添加氧化锆可以提高陶瓷的常温抗压强度，特别是对于熔融氧化镁-氧化锆共混物体系，并能改善抗侵蚀性[31]。

In the presented study, MgO and **MgO-ZrO_2 refractory ceramics** made from secondary magnesia clinker obtained from fired spent magnesia-carbon refractories were compared with materials obtained in a traditional way—from commercial raw materials. It has been found that spent MgO-C refractories can be successfully recycled, but it is worth bearing in mind that the obtained secondary magnesia clinker is characterized by lower quality. This quality can be considerably improved by adding zirconia and/or producing fused MgO-ZrO_2 co-clinker, which reduces open porosity and increases bulk density. Moreover, zirconia addition increases the cold crushing strength of ceramics, especially in the case of fused magnesia-zirconia co-clinker, and improves corrosion resistance[31].

将通过共沉淀法制备并在 900 ℃下煅烧 1 h 后得到的纳米晶镁铝（MA）尖晶石粉末以质量分数为 0~25%加入**氧化镁-氧化锆复合材料**中，并在 1600 ℃下烧结 2 h。使用扫描电子显微镜（SEM）和 X 射线衍射（XRD）技术研究烧结后复合材料的微观结构和相组成。还研究了烧结复合材料的体积密度、表观孔隙率、体积收缩率和杨氏模量。结果显示，添加质量分数高达 20%的纳米尖晶石能够增加复合材料的烧结能力和杨氏模量。显微结构显示，纳米尖晶石和氧化锆在氧化镁晶粒三角区的存在填补了陶瓷基体的间隙，增强了复合材料的密实度[32]。

Nanocrystalline magnesium aluminate (MA) spinel powder produced through a coprecipitation method and calcined at 900 ℃ for 1 h was added to **magnesia-zirconia composite** in the range of 0-25% (mass fraction) and sintered at 1600 ℃ for 2 h. Scanning electron microscope (SEM) and X-ray diffraction (XRD) techniques were used for studying the microstructure and the phase composition of the sintered composites. Bulk density, apparent porosity, volume shrinkage, and Young's modulus of the sintered composites were also investigated. The results revealed that the nanospinel addition up to 20% (mass fraction) increases the sintering ability and Young's modulus of the composite bodies. Microstructure showed that the presence of nanospinel and zirconia in the triple point between magnesia grains closed the gaps in the ceramic matrix and enhanced the compactness of the composites[32].

参考文献

[1] Yang Jie, Lu Shaowen, Wang Liangyong. Fused magnesia manufacturing process: A survey [J]. Journal of Intelligent Manufacturing, 2020, 31 (2): 327-350.

[2] Liu Zhuangzhuang, Wang Shuai, Huang Jian, et al. Experimental investigation on the properties and microstructure of magnesium oxychloride cement prepared with caustic magnesite and dolomite [J]. Construction and Building Materials, 2015, 85: 247-255.

[3] Unterreiter G, Kreuzer D R, Lorenzoni B, et al. Compressive creep measurements of fired magnesia bricks at elevated temperatures including creep law parameter identification and evaluation by finite element analysis [J]. Ceramics, 2020, 3 (2): 210-222.

[4] Tang Huimin, Peng Zhiwei, Gu Foquan, et al. Alumina-enhanced valorization of ferronickel slag into refractory materials under microwave irradiation [J]. Ceramics International, 2020, 46 (5): 6828-6837.

[5] Hou Qingdong, Luo Xudong, Xie Zhipeng, et al. Preparation and characterization of microporous magnesia-based refractory [J]. International Journal of Applied Ceramic Technology, 2020, 17 (6): 2629-2637.

[6] Zhang Yakun, Sun Luen, Lei Yun, et al. Corrosion behavior of carbon, Al_2O_3, and MgO refractories during the preparation of a Ti-Si-Al alloy via the aluminothermic reduction of a Ti-bearing blast-furnace slag [J]. Ceramics International, 2021, 47 (13): 18044-18052.

[7] Wolters D. Testing in reducing and non-oxidizing atmosphere [J]. Industrie Ceramique, 1982, 766: 805-808.

[8] Ghasemi-Kahrizsangi S, Karamian E, Ghasemi-Kahrizsangi A, et al. The impact of trivalent oxide nanoparticles on the microstructure and performance of magnesite-dolomite refractory bricks [J]. Materials Chemistry and Physics, 2017, 193: 413-420.

[9] Li Xiaohui, Gu Huazhi, Huang Ao, et al. Bonding mechanism and performance of rectorite/ball clay bonded unfired high alumina bricks [J]. Ceramics International, 2021, 47 (8): 10749-10763.

[10] Wen Tianpeng, Jin Yao, Liu Zhaoyang, et al. Preparation of alumina-rich calcium aluminate refractories with improved sintering densification and properties [J]. Journal of the Australian Ceramic Society, 2023.

[11] Wang Yulong, Li Xiangcheng, Zhu Boquan, et al. Microstructure evolution during the heating process and its effect on the elastic properties of CAC-bonded alumina castables [J]. Ceramics International, 2016, 42 (9): 11355-11362.

[12] Cui Yuanyuan, Zhong Kai, Zhu Shaojun, et al. Research and application of new non-burned periclase-magnesia alumina spinel bricks [J]. China's Refractories, 2019, 28 (2): 18-22.

[13] Pacheco G R C, Gonçalves G E, Lins V F C. Design of magnesia-spinel bricks for improved coating adherence in cement rotary kilns [J]. Ceramics, 2021, 4 (4): 652-666.

[14] Chen Wei, Zhang Xiaohui, Zhang Haijun. Slag resistance of refractories for direct reduction of laterite nickel ores in rotary kilns [J]. China's Refractories, 2020, 29 (2): 11-20.

[15] Xiao Junli, Chen Junfeng, Li Yuanyuan, et al. Corrosion mechanism of cement-bonded Al_2O_3-$MgAl_2O_4$ pre-cast castables in contact with molten steel and slag [J]. Ceramics International, 2022, 48 (4): 5168-5173.

[16] Klischat H J, Wirsing H. Higher thermochemical resistance by installation of Magnesia Forsterite bricks [C] //Proceedings of the Unified International Technical Conference on Refractories (UNITECR 2013). John Wiley & Sons, 2014: 193.

[17] Kullatham S, Sirisoam T, Lawanwadeekul S, et al. Forsterite refractory brick produced by talc and magnesite from Thailand [J]. Ceramics International, 2022, 48 (20): 30272-30281.

[18] Khattab R M, Wahsh M M S, Khalil N M. Ceramic compositions based on nano forsterite/nano magnesium aluminate spinel powders [J]. Materials Chemistry and Physics, 2015, 166: 82-86.

[19] Tavangarian F, Emadi R. Synthesis and characterization of spinel-forsterite nanocomposites [J]. Ceramics International, 2011, 37 (7): 2543-2548.

[20] Meng Chao, Meng Qingxin, Li Xiaolong, et al. Preparation and application of light-weight raw materials using natural forsterite [J]. China's Refractories, 2021, 30 (3): 37-42.

[21] Chen Ding, Gu Huazhi, Huang Ao, et al. Mechanical strength and thermal conductivity of modified expanded vermiculite/forsterite composite materials [J]. Journal of Materials Engineering and Performance, 2016, 25: 15-19.

[22] Liu Gengfu, Li Yawei, Zhu Tianbin, et al. Influence of the atmosphere on the mechanical properties and slag resistance of magnesia-chrome bricks [J]. Ceramics International, 2020, 46 (8): 11225-11231.

[23] Liu Wei, Wang Ling, Ma Baozhong, et al. Reactions between magnesia-chrome refractories and copper converter slags: Corrosion behavior and prevention by Fe-rich layer formation [J]. Ceramics International, 2022, 48 (10): 14813-14824.

[24] Petkov V, Jones P T, Boydens E, et al. Chemical corrosion mechanisms of magnesia-chromite and chrome-free refractory bricks by copper metal and anode slag [J]. Journal of the European Ceramic Society, 2007, 27 (6): 2433-2444.

[25] Szczerba J, Madej D, Czapka Z. The impact of work environment on chemical and phase composition changes of magnesia-spinel refractories used as refractory lining in secondary metallurgy device [C] //IOP Conference Series: Materials Science and Engineering. IOP Publishing, 2013, 47 (1): 012020.

[26] Jin Peng, Zhao Hongbo, Zhang Lixin, et al. Review of basic refractories and development history of Sinosteel Luonai Basic Refractories [J]. China's Refractories, 2022, 31 (1): 45-52.

[27] Jones P T, Blanpain B, Wollants P, et al. Degradation mechanisms of magnesia-chromite refractories in vacumm-oxygen decarburisation ladles during production of stainless steel [J]. Ironmaking & steelmaking, 2000, 27 (3): 228-237.

[28] 镁铬砖的制备工艺参数与抗 RH 炉渣侵蚀的相关性 [J]. 北京科技大学学报, 2009, 31 (2): 207-214.

[29] Zhang Huifang, Huang Hongliang, Zhang Lifang, et al. Study on properties of re-bonded magnesium zirconium refractory materials and materials application [C] //Advanced Materials Research. Trans Tech Publications Ltd, 2012, 578: 95-98.

[30] Guo Zongqi, Lei Zhongxing, Ma Ying, et al. Bonding phase formation in eco-friendly periclase-spinel-Al bricks used in RH degassing process [J]. Journal of the European Ceramic Society, 2023, 43 (6): 2663-2674.

[31] Guo Zongqi, Ma Ying, Dang Xiaomei, et al. Critical whisker bond of a profound chromium-free refractory for RH degassers [J]. Sādhanā, 2022, 47 (2): 70.

[32] Khattab R M, Wahsh M M S, Khalil N M, et al. Effect of nanospinel additions on the sintering of magnesia-zirconia ceramic composites [J]. ACS applied materials & interfaces, 2014, 6 (5): 3320-3324.

4 中性耐火材料制品

4.1 石墨

4.1.1 术语词组

石墨（质）制品 graphite articles, graphite products
石墨（质）耐火材料 graphite refractory
石墨相氮化碳 graphitic carbon nitride, C_3N_4
石墨模型 graphite mould
石墨填料 graphite packing
石墨纤维 graphite fibers
石墨压模 graphite compression mould
石墨电极 graphite electrode
石墨阳极 graphite anode; graphite cathode
石墨质熔池 graphite bath
石墨舟 graphite boat
石墨坩埚 graphite dust; graphite powder; powdered graphite; graphite crucible
石墨风口砖 graphite tuyere block
石墨棒 graphite bar
石墨板 graphite sheet
石墨粉 graphite powder
石墨管 graphite tube
石墨环 graphite annulus
酸化石墨 acidified graphite
多孔石墨 porous graphite
膨胀石墨 expanded graphite; exfoliated graphite; EG
氟化石墨 graphite fluoride
微晶石墨 microcrystalline graphite
鳞片石墨 flake graphite
石墨火泥 graphite mortar
石墨黏土砖 carbon-clay brick
石墨大理石 graphite marble
石墨发热体 graphite heater

4.1.2 术语句子

石墨层中的碳原子相互连接形成碳环。
In **graphite** sheets, carbon atoms bond together in rings.

碳的同素异形包括金刚石、**石墨**和木炭，这些物质都具有相同的化学性质。

The allotropes of carbon include diamond, **graphite** and charcoal, all with the same chemical properties.

钢铁工业消耗了全世界耐火材料总产量的近50%，其中很大一部分是**含石墨的耐火材料**。

The steel industry consumes almost 50% of the total produced refractory materials in the world and a large proportion of this consumption is **graphite containing refractories**.

石墨制品是重要的战略资源，用途十分广泛。

Graphite products are important strategic resources and are widely used.

石墨坩埚主要用于金属材料的熔融，它分天然石墨质和人造石墨质两类。

Graphite crucible is mainly used for melting metal materials. It is divided into natural graphite and artificial graphite.

石墨舟有三个储液器和一个可以滑动的石墨滑板。

This **graphite boat** is provided with three reservoirs and a movable graphite slide.

大多数浸入式进口喷嘴由**氧化物石墨质耐火材料**制成。

Most submerged entry nozzles are fabricated from **oxide-graphite refractory**.

锆石墨质耐火材料是一种复合材料，由二氧化锆颗粒和鳞片石墨组成。

Zircon graphite refractory material is a composite material composed of zirconia particles and flake graphite.

随着**微晶石墨**加入量的增加，试样的体积密度略有升高，常温抗折强度和耐压强度稍有提高，高温抗折强度有所增加，抗热震性稍有提高，抗渣侵蚀性有所下降。

With the increasing addition of **microcrystalline graphite**, the bulk density of the specimens increases slightly, the cold modulus of rupture and cold crushing strength increase slightly, the hot modulus of rupture and thermal shock resistance increase slightly, but the slag corrosion resistance decreases.

氧化锆-石墨耐火材料通常用于连铸用浸入式水口的渣线部位，比 MgO-C 和 Al_2O_3-C 耐火材料具有优良的抗热震性和抗侵蚀性。

Zirconia graphite refractory materials are commonly used in the slag line of submerged nozzles for continuous casting, and have excellent thermal shock resistance and corrosion resistance compared to MgO-C and Al_2O_3-C refractory materials.

4.1.3 术语段落

结果表明，当**石墨坩埚**厚度为 20 mm 时，可获得良好的对流形态、平坦的固液界面、合理的 V/G 值等，有利于节约多晶硅的生产成本并提高多晶硅的品质，为生产实践中工艺方案优化及缺陷分析等提供重要的理论依据[1]。

The results show that when the thickness of the **graphite crucible** is 20 mm, good convection morphology, flat solid-liquid interface, reasonable V/G value, etc. can be obtained, which are beneficial to save the production cost of polysilicon and improve the quality of polysilicon. It provides an important theoretical basis for optimizing process plans and analyzing defects in production practice[1].

以生物质燃烧剂作为热源时，高温氧蚀和碱金属腐蚀之间的耦合作用是导致**碳化硅石墨坩埚**损坏的主要原因。其中，碱金属与 SiO_2 反应形成低熔点化合物并被不断剥蚀，从而导致碳化硅石墨坩埚的腐蚀程度加剧[2]。

When the biomass burner is used as the heat source, the coupling effect between high-temperature oxygen corrosion and alkali metal corrosion is the main reason for the damage of **SiC graphite crucible**. Among them, the alkali metal reacts with SiO_2 to form low melting point compounds and is continuously denuded, which leads to the aggravation of the corrosion of the silicon carbide graphite crucible[2].

由于**石墨**与渣的不润湿性，能够阻止渣向试样内部渗透，所以，在坩埚底部与氧气隔绝的情况下，石墨的抗侵蚀性能得以体现；但石墨氧化后材料结构变得疏松多孔，渣通过石墨氧化后留下的通道向材料内部渗透，加剧了侵蚀[3]。

Because the non-wettability of **graphite** and slag can prevent the slag from penetrating into the sample, the corrosion resistance of graphite can be reflected when the bottom of the crucible is isolated from oxygen; However, after graphite oxidation, the material structure becomes loose and porous, and the slag penetrates into the material through the channels left by graphite oxidation, which intensifies the erosion[3].

当热处理温度由 1100 ℃升高至 1200 ℃，**膨胀石墨**表面原位生成 β-SiC 晶须数量逐渐增多，长径比增加；当热处理温度为 1300 ℃时，膨胀石墨自身发生结构蚀变形成碳化硅晶须，同时其表面生成了长径比更大的 β-SiC 晶须和少量的方石英、β-Si_3N_4 及 Si_2N_2O 等陶瓷相。在铝碳材料中引入硅修饰膨胀石墨时，其对材料力学性能产生显著的影响。对于 800～1000 ℃处理的铝碳材料，引入低温处理（1100～1200 ℃）的硅修饰膨胀石墨能够显著提高材料的力学性能；而对于 1200～1400 ℃处理的铝碳材料，引入硅修饰膨胀石墨，不利于改善铝碳耐火材料的力学性能[4]。

β-SiC whiskers in-situ formed on the surface of **expanded graphite**. And their amount and length / diameter ratio increased with the firing temperature ranging from 1100 ℃ to 1200 ℃. At 1300 ℃, large amount of β-SiC whiskers with larger length / diameter ratio formed at the cost of the collapse of expanded graphite; Meantime a little of cristobalite, β-Si_3N_4, Si_2N_2O formed on the surface of expanded graphite. The SEG addition had obvious effect on the mechanical properties of Al_2O_3-C refractories. Furthermore, mechanical properties of Al_2O_3-C refractories coked at 800-1000 ℃ are improved obviously with the addition of SEG synthesized at low temperatures (1100 ℃ to 1200 ℃). However, the addition of SEG had negative effect on the mechanical properties of Al_2O_3-C refractories coked at 1200-1400 ℃[4].

石墨，用作传统镁碳（MgO-C）耐火材料中的碳源，用无机前体涂覆以防止石墨与氧之间的接触。与常见的抗氧化剂相比，由此产生的涂层可以有效地保护石墨免受氧气的伤害。涂层试剂由硅酸盐和金属醇盐组成，通常称为无机前体，对石墨表面进行改性。改性石墨在 1000 ℃的环境气氛中表现出优异的稳定性，而传统 MgO-C 耐火材料中的石墨与氧反应并表现出显著的质量损失。涂层试剂均匀地涂覆在石墨表面，它在防止石墨氧化方面非常有效。此外，制备的耐火材料样品的形状保持固定，即使当酚醛树脂的含量较低时也是如此。因此，通过在石墨上涂覆无机前体，成功地制备了具有高抗氧化性的生态友好型 MgO-C 耐火材料[5]。

Graphite, used as a carbon source in a conventional magnesia-carbon (MgO-C) refractory, was coated with an inorganic precursor to prevent the contact between graphite and oxygen. The coating layer thus generated and could effectively protect graphite from oxygen, compared to common antioxidants. The coating reagent was composed of silicate and metal alkoxide, generally called the inorganic precursor, which modified the graphite surface. The modified graphite showed excellent stability at 1000 ℃ in an ambient atmosphere, whereas the graphite in a conventional MgO-C refractory reacted with oxygen and showed a significant weight loss. The coating reagent was coated uniformly on the graphite surface and it was very efficient in preventing the oxidation of graphite. Furthermore, the shape of prepared refractory samples remained fixed, even when the content of phenol resin was low. Consequently, ecofriendly MgO-C refractories with high oxidation resistance were successfully prepared by the coating of an inorganic precursor onto graphite[5].

膨胀石墨是以天然鳞片石墨为原料，经氧化、插层、洗涤、干燥、加热膨胀等工艺制成疏松多孔结构的石墨。它已广泛应用于垫片、吸附、电磁干扰屏蔽、电化学应用、应力传感和绝缘体。这种石墨薄片有中间和边缘两个反应区。在强氧化条件下，石墨薄片的边缘和层分别被氧化为插层[6]。

Exfoliated graphite with a loose and porous structure is made from natural flake graphite by the process of oxidation, intercalating, washing, drying and heating expansion. It has been widely used in gasket, adsorption, electromagnetic interference shielding, electrochemical applications, stress sensing and thermal insulator. The graphite flake has two reactive regions, such as the middle and the edge. Under a strong oxidizing condition, the edge and the layer of graphite flake has been respectively oxidized to intercalation[6].

4.2 含碳耐火材料制品

4.2.1 术语词组

含碳耐火材料 carbon-bearing refractory; carbon containing refractory
低碳耐火材料 low carbon refractory
碳复合耐火材料 carbon composite refractory
碳化物耐火材料 carbide refractory
碳质黏土 carbonaceous clay
碳棒 carbon bar
碳砖（碳素砖）carbon brick
微孔碳砖 micro-pore carbon brick
焙烧碳砖 roasted carbon brick
自焙碳砖 self-baking carbon brick
预焙碳块 prebaked carbon block
高炉用碳砖 carbon brick for blast furnace
高温模压碳砖 high-temperature mould pressing carbon brick
电炉碳砖 electric furnace carbon brick
碳砖糊料 carbon brick paste
半石墨碳砖 semi graphite carbon brick

碳纤维 carbon fiber; carbon filament
碳素捣打料 carbon ramming mix
碳质内衬 carbon lining
碳质泥料 carbon loam; carbon mass; carbon mix
碳质火泥 carbon mortar
镁碳砖 magnesia carbon brick; MgO-C brick
低碳镁碳砖 low carbon magnesia carbon brick
镁钙碳砖 magnesia calcia carbon brick
铝碳质耐火材料 alumina carbon refractory
铝碳砖 alumina carbon brick; Al_2O_3-C brick
铝锆碳质耐火材料 alumina zirconia carbon refractory
铝锆碳砖 Al_2O_3-ZrO_2-C brick
铝镁碳质耐火材料 alumina magnesia carbon refractory
Al_2O_3-SiC-C 砖 alumina-SiC-C brick

4.2.2 术语句子

碳棒耐高温，导电性好，不易断裂。

Carbon rod is resistant to high temperature, has good conductivity and is not easy to break.

碳砖是以碳质材料为原料，加入适量结合剂制成的耐高温中性耐火材料制品。

Carbon brick is a high-temperature neutral refractory product made of carbonaceous materials and a proper amount of binder.

自焙炭砖是我国炼铁和炭素生产工作者经过多年的艰苦探索开发出的一种独创的炭质耐火材料。

Self-baking carbon brick is a kind of original carbon refractory material developed by iron smelting and carbon production workers in China after years of hard exploration.

使用沥青作结合剂制备的**微孔碳砖**残碳量较高。

The carbon residue of **micro-pore carbon bricks** prepared with asphalt as binder is high.

含碳耐火材料在全世界都得到了广泛的普及和应用，其典型代表有镁碳砖、铝碳砖。

Carbon containing refractories have been widely popularized and applied all over the world, and their typical representatives are magnesium carbon bricks and aluminum carbon bricks.

结果表明，随着保护渣黏度的降低或渣中［F^-］含量的增高，ZrO_2-**石墨质**

耐火材料蚀损严重。

Results showed that while viscosity number of mould powder was reduced or [F^-] content were enlarged, ZrO_2-**graphite refractory** was badly corroded.

由于自身优异的综合性能，**铝碳耐火材料**广泛应用于钢铁连铸领域。

Because of its excellent comprehensive performance, **aluminum carbon refractory** materials are widely used in the field of steel continuous casting.

在 Al 复合 **Al_2O_3-C 材料**中加入 Zn 粉可以显著改善材料的抗氧化性。

Adding Zn powder into Al Composite **Al_2O_3-C material** can significantly improve the oxidation resistance of the material.

MgO-C 耐火材料作为碳复合材料中最重要的品种之一被广泛运用于炼钢行业。

As one of the most important varieties of carbon composite materials, **MgO-C refractories** are widely used in the steelmaking industry.

在 **MgO-C 材料**中加入氧化锆和锆英石后，试样的高温抗折强度均随加入量的增加而降低，其下降幅度大致随添加量的增加而增大。

After adding ZrO_2 and zircon into **MgO-C material**, the high-temperature flexural strength of the sample decreases with the increase of the amount of addition, and the decreasing range increases with the increase of the amount of addition.

4.2.3 术语段落

根据**含碳耐火材料**的氧化损毁机理，添加抗氧化剂法依旧是含碳耐火材料最常用的防氧化技术。金属抗氧化剂除了生成金属氧化物和碳化物阻止含碳耐火材料的氧化外，通过固相反应生成的陶瓷相产物还可以提高含碳耐火材料的力学性能和抗渣侵蚀性能，过渡金属和金属合金作为抗氧化剂还可以催化热解碳石墨化以及促进碳化物晶须的生成[7]。

According to the oxidative damage mechanism of **carbon-containing refractories**, the addition of antioxidant is still the most commonly used oxidation resistant technology for carbon-containing refractories. In addition to the formation of metal oxides and carbides to prevent the oxidation of carbon-containing refractory materials, metal antioxidants can also improve the mechanical properties and slag corrosion resistance of carbon-containing refractories by solid state reaction. The transition metals and metal alloys. As an antioxidant, it also has catalytic pyrolytic carbon graphiti-zation and promotes the formation of carbide whiskers[7].

碳纤维改性树脂经 220 ℃热处理后，树脂碳与碳纤维间形成紧密结构；经 1100 ℃、1400 ℃热处理后碳纤维与树脂碳之间较为疏松；通过 SEM 结果发现，采用将碳纤维与细粉湿混的工艺将碳纤维引入**铝碳耐火材料**中，碳纤维均匀分散于基质中，铝碳耐火材料力学性能最优；碳纤维复合铝碳耐火材料经 220 ℃、1100 ℃热处理后碳纤维表面变化不大，经 1400 ℃热处理后碳纤维表面形成晶须，可提高材料的力学性能[8]。

After heat treatment at 220 ℃, a compact structure was formed between resin carbon and carbon fiber. After heat treatment at 1100℃ and 1400 ℃, the carbon fiber and resin carbon were relatively loose. Through SEM results, it was found that the mechanical properties of **aluminum-carbon refractories** were the best when carbon fibers were uniformly dispersed in the matrix by wet mixing process of carbon fibers and fine powder. After heat treatment at 220 ℃ and 1100 ℃, the surface of carbon fiber had little change. After heat treatment at 1400 ℃, the surface of carbon fiber had whisker, which could improve the mechanical properties of the material[8].

Al_2O_3-C 材料被广泛应用于长水口、整体塞棒等，要求这些制品具有良好的抗热震性，高碳含量是良好抗热震性的基本保障。但是，高碳含量又不可避免地导致制品在使用过程中向钢水增碳、降低钢水质量[9]。

Al_2O_3-C materials are widely used in long nozzle, integral stopper, etc. These products are required to have good thermal shock resistance, and high carbon content is the basic guarantee of good thermal shock resistance. However, high carbon content inevitably leads to carburization in molten steel and the reduction of molten steel quality during the use of products[9].

将碳含量保持在 4%的水平，研究改变纳米碳源和石墨含量之比对通过压制、干燥和烧成制备的**铝碳耐火材料**试样的体积密度和耐压强度、不同温度下试样的物相组成、微观结构变化以及对抗氧化性能的影响。结果表明：由于纳米碳源的颗粒尺寸较小，可以有效填充其他耐火物料之间的间隙，从而降低了孔隙率并提高了强度。此外，材料中的纳米碳源促进了碳化铝的原位生成，也显著改善了材料的强度[10]。

Maintain the carbon content at a level of 4%, and study the volume density and compressive strength of **aluminum carbon refractory** samples prepared by pressing, drying, and firing, as well as the phase composition and microstructure changes of the samples at different temperatures, as well as the impact of changing the ratio of nano carbon source to graphite content. The results indicate that due to the small particle size of the nano carbon source, it can effectively fill the gaps between other refractory materials, thereby reducing porosity and improving strength. In addition, the nano carbon source in the material promotes the in-situ generation of aluminum carbide and significantly improves its strength[10].

含碳耐火材料具有优异的抗热震性及抗渣侵蚀性等性能，被广泛应用于炼钢系统的关键部位。但这些耐火材料中过高的碳含量会产生钢液增碳、能耗增大等问题。为了满足含碳耐火材料低碳化发展的要求，高效抗氧化剂的研发至关重要[11]。

Carbon-containing refractories are widely used in key parts of steelmaking systems because of their excellent thermal shock resistance and slag resistance. However, the increase of energy consumption. To satisfy the requirements of their low-carbon development, the research on high-efficiency antioxidants is of vital importance[11].

镁碳质耐火材料具有耐火度高、抗热震性强、高温蠕变小、抗侵蚀性和抗剥落性优良等性能，因此被广泛应用于氧气转炉炉衬、电炉炉墙和钢包渣线等部位。然而由于碳容易被氧化而形成脱碳层，导致材料组织结构疏松，强度降低。在熔渣侵蚀和机械冲刷等作用下，氧化镁颗粒逐渐被熔蚀、脱落而损毁。针对以上问题，本工作研究了结合剂种类、工业废料 Si-SiC 复合粉体的种类及含量和微晶石墨含量对低碳镁碳砖性能及显微结构的影响，期望提高低碳镁碳耐火材料的抗氧化性，并降低低碳耐火材料的成本[12]。

MgO-C refractories are widely used in the steel making industries such as BOF, EAF and steel ladles, because of their high refractoriness, excellent thermal shock, low high temperature creep and excellent corrosion resistance and antistrip performance. However, the main drawback of MgO-C refractories is its poor oxidation resistance. After the oxidation of carbon, the structure of MgO-C brick is destroyed, and then the intensity reduced. Under the effect of slag erosion and mechanical erosion, MgO particles gradually damaged by corrosion or fell off. The research mainly studied the effects of binder types, the kinds and contents of waste Si-Si C composite powder and the contents of amorphous graphite on properties and microstructure of low-carbon MgO-

C refractory, and reduce the cost of MgO-C refractory[12].

在 **MgO-C 材料**中加入或原位生成非氧化物相可有效提高材料的高温力学性能和抗热震性。郑州大学高温材料研究所与北京首钢股份有限公司合作，采用“金属复合、原位反应生成非氧化物结合”的创新工艺研制了 **Al/Si 复合低碳 MgO-C 渣线砖**。与传统 MgO-C 钢包渣线砖相比，它们的碳含量大大降低，抗热震性和抗渣性水平相当，高温抗折强度和抗氧化性更好[13]。

The addition or in-situ formation of non-oxide phase in **MgO-C material** can effectively improve the high temperature mechanical properties and thermal shock resistance of the material. In cooperation with Beijing Shougang Co. , Ltd. , the Institute of High Temperature Materials of Zhengzhou University has developed **Al/Si composite low carbon MgO-C slag line bricks** by using the innovative process of "metal composite and in-situ reaction to produce non oxide bonding". Compared with traditional MgO-C ladle slag line bricks, their carbon content is greatly reduced, their thermal shock resistance and slag resistance are equivalent, and their high temperature bending strength and oxidation resistance are better[13].

考虑到**含碳耐火材料**中需要用到大量的石墨为原料以及采用高黏度酚醛树脂为结合剂的技术特征，在借鉴当前国际上利用三辊研磨制备石墨/膨胀石墨与环氧树脂复合材料等新技术的基础上，本研究创新性地提出通过三辊差速剥离天然鳞片石墨的方法来获得高纵/横比、均匀分散在酚醛树脂中的微/纳米石墨片。并进一步开发制备新型 **MgO/Al_2O_3-C 低碳含碳耐火材料**，旨在保证耐火材料综合性能的前提下大幅降低碳含量[14]。

Considering the technical characteristics of using a large amount of graphite as raw material and high viscosity phenolic resin as binder in **carbon containing refractory materials**, and drawing on new international technologies such as using three roller grinding to prepare graphite/expanded graphite and epoxy resin composite materials, this study innovatively proposes the method of using three roller differential stripping of natural flake graphite to obtain high longitudinal/transverse ratio. Further development and preparation of new **MgO/Al_2O_3-C low-carbon carbon containing refractory materials** using micro/nano graphite sheets uniformly dispersed in phenolic resin, aiming to significantly reduce carbon content while ensuring the comprehensive performance of the refractory material[14].

在 **MgO-C/Al_2O_3-C 复合功能耐火材料**中，石墨阻碍了 MgO 与 Al_2O_3 的直接

接触。Mg（g）是由氧化镁和石墨的碳热反应生成的。在 MgO-C 与 Al_2O_3-C 材料的界面处，Mg（g）从 MgO-C 材料扩散到 Al_2O_3-C 材料中，并在 1550 ℃时与 Al_2O_3 反应生成 $MgAl_2O_4$（MA）。界面处 MA 的形成削弱了复合材料的结合强度，降低了复合材料的力学性能。MA 的加入并不能减弱界面传质和界面尖晶石化对样品性能的不利影响[15]。

In **MgO-C/Al_2O_3-C composite functional refractory** component, graphite blocks the direct contact between MgO and Al_2O_3. Mg (g) is formed by carbothermal reaction of MgO and graphite. At the interface between MgO-C and Al_2O_3-C material, Mg (g) will diffuse from MgO-C material into Al_2O_3-C material and react with Al_2O_3 to form $MgAl_2O_4$ (MA) at 1550 ℃. The formation of MA at the interface weakens the bonding strength and decreases the mechanical properties of composite component samples. The addition of MA cannot weaken the adverse effect of interface mass transfer and interfacial spinellisation on the properties of samples[15].

4.3 SiC 及含 SiC 耐火材料制品

4.3.1 术语词组

碳化硅纤维 silicon carbide fiber
碳化硅晶须 silicon carbide whiskers
碳化硅纳米线 silicon carbide nanowires
碳化硅纳米管 silicon carbide nanotubes
碳化硅纳米片 silicon carbide nanosheets
碳化硅膜 silicon carbide film
单晶碳化硅 single crystal silicon carbide
多晶碳化硅 polycrystalline silicon carbide
非晶碳化硅 amorphous silicon carbide
泡沫碳化硅 foam silicon carbide
多孔碳化硅陶瓷 porous silicon carbide ceramics
立方碳化硅 cubic silicon carbide
氧化物结合 SiC 制品 oxide bonded SiC
黏土结合碳化硅制品 clay bonded silicon carbide products
莫来石结合碳化硅制品 mullite bonded silicon carbide products
二氧化硅结合碳化硅制品 silicon dioxide bonded silicon carbide products
氮化物结合 SiC 制品 nitride bonded SiC
氮化硅结合碳化硅制品 silicon nitride bonded silicon carbide products
塞隆结合碳化硅制品 Sialon bonded silicon carbide products
氧氮化硅和复相氮化物结合碳化硅制品 silicon oxy nitride and compound phase nitride bonded silicon carbide products
自结合 SiC 制品 self-bonded silicon carbide products
β-SiC 结合 SiC 制品 β-SiC bonded silicon carbide products

重结晶 SiC 制品 recrystallized silicon carbide products
渗硅反应 SiC 质制品 silicon infiltration reaction silicon carbide products
半 SiC 质制品 semi-silicon carbide products
碳化硅陶瓷基复合材料 silicon carbide ceramic matrix composite material

4.3.2 术语句子

碳化硅通常与碳纤维混合，用于非氧化物基体材料中，从而获得增强复合材料，在 2600 ℃以上的温度下进行应用。

SiC is usually mixed with carbon fiber and used in non-oxide matrix materials to obtain reinforced composites that can be used at the temperature above 2600 ℃.

纤维增强 SiC 陶瓷基复合材料具有密度低、强度高、耐高温、抗氧化、耐腐蚀等优点，在航空航天及其他高温条件使用领域具有广泛的应用潜力。

Fiber reinforced SiC matrix composites (FRCMC-SiC) have significant potential in aerospace application and other high temperature conditions due to their special properties such as low density, high strength, thermostability, anti-oxidation and anti-ablation.

引入原位 SiC 纳米线可以显著提升 **SiC 陶瓷**的抗热震性能。

The introduction of in-situ SiC nanowires can significantly improve the thermal shock resistance of **SiC ceramics**.

陶瓷黏结剂含量的增加使**碳化硅多孔陶瓷**的气孔率快速下降。

The porosity of **SiC porous ceramics** decreased with increasing the ceramic binder contents.

氧化物结合碳化硅制品的结合相矿物组成主要为石英、莫来石和硅酸盐玻璃相。

The mineral composition of **oxide-bound silicon carbide products** is mainly quartz, mollite and silicate glass phase.

黏土结合 SiC 制品主要用作各种工业窑炉的隔焰板、陶瓷窑具（棚板、立柱、匣体等）。

Clay bonded silicon carbide products are mainly used as various industrial kiln flame partition plate, ceramic kiln tools (shed plate, column, box body, etc.).

莫来石结合 SiC 制品使用性能明显优于黏土结合 SiC 制品，目前在陶瓷、有色冶金、机械等行业仍在使用。

Mullite bonded silicon carbide products are obviously better than clay bonded silicon carbide products, and are still used in ceramics, non-ferrous metallurgy, machinery and other industries.

氮化物结合碳化硅制品是指以 Si_3N_4、Sialon、Si_2N_2O 和 AlN 等单相或复相氮化物为结合相的 SiC 质高级耐火材料。

Nitride bonded silicon carbide products refer to the SiC-quality advanced refractory material with mono-phase or complex-phase nitride such as Si_3N_4, Sialon, Si_2N_2O, and AlN as the binding phase.

氮化物结合 SiC 制品秉承了氮化物和 SiC 材料的许多优异性能。

Nitride bonded silicon carbide products adhere to the many excellent properties of nitride and SiC materials.

在大多数 **Si_3N_4 结合 SiC 产品**中，结合相 Si_3N_4 通常以 α-Si_3N_4 为主。

In most **Si_3N_4 bonded silicon carbide products**, the binding phase Si_3N_4 is usually dominated by α-Si_3N_4.

Sialon 结合 SiC 制品主要采用反应烧结方法制备，其氮化烧成温度通常比 Si_3N_4 结合 SiC 高，综合成本略高于 Si_3N_4 结合 SiC 产品。

Sialon bonded silicon carbide products are mainly prepared by reaction sintering method. The nitride firing temperature is usually higher than Si_3N_4 bonded silicon carbide products, and the comprehensive cost is slightly higher than Si_3N_4 bonded silicon carbide products.

β-SiC 结合 SiC 砖可用作高炉衬砖和风口组合砖、垃圾焚烧炉内衬等。

β-SiC bonded silicon carbide brick can be used as high furnace lining brick and air outlet combination brick, waste incinerator lining, etc.

重结晶 SiC 制品（R-SiC）是一种无其他结合相的 SiC 制品，它是一种靠 SiC 晶粒的再结晶作用而形成的晶粒与晶粒直接相连的 α-SiC 陶瓷单相材料。

Recrystallized silicon carbide products (R-SiC) is a kind of SiC product without any other binding phase, it is a α-SiC ceramic monophase material formed by the recrystallization of SiC grains.

渗硅碳化硅制品（SiSiC）是一种性能优良的 SiC/Si 复相陶瓷材料，材料中 Si 形成连续相或者 Si 和 SiC 互为连续相，主要采用反应烧结工艺制备。

Silicon infiltration reaction silicon carbide products (SiSiC) is an excellent SiC / Si complex phase ceramic material, where Si forms a continuous phase or Si and SiC are a continuous phase of each other, mainly prepared by reaction sintering process.

氧化脆化是 **SiC/SiC 复合材料**在中温范围内的重要失效机制，材料在高温下的失效主要是由纤维强度退化、蠕变及界面氧化引起的。

Oxidative embrittlement is a crucial failure mechanism of **SiC/SiC composites** in the medium temperature range. While the failure of the composites at high temperatures is mainly caused by fiber strength degradation, fiber creep and interface oxidation.

4.3.3 术语段落

根据氧分压将 **SiC 纤维**的氧化行为分为被动氧化与主动氧化；氧化环境如水氧环境下的高温氧化加快了 SiC 纤维氧化层厚度的增加和裂纹的产生，致使 SiC 纤维的力学性能下降；SiC 纤维的自身组成与微观结构也会对氧化结果产生各不相同的影响[16]。

The oxidation behavior of **SiC fibers** is divided into passive and active oxidation according to the oxygen partial pressure. The high temperature oxidation in the water-oxygen environment accelerates the increase of the SiC fibers oxide layer thickness and the generation of cracks, resulting in the degradation of the mechanical properties. The composition and microstructure also have different effects on the oxidation results[16].

随着科技的进步，为了满足不同领域的需求，必然会对多孔 SiC 的结构产生许多详细特定的要求。**多孔 SiC 陶瓷**的制备将朝显微结构精细化，气孔率、气孔形貌及力学性能可控方向发展，以满足不同领域对多孔 SiC 陶瓷的需求[17]。

With the progress of science and technology, many specific requirements for the structure of porous SiC in order to meet the requirements of different fields. The preparation of **porous SiC ceramics** will refine towards the microstructure, and control the stomatal rate, stomatal morphology and mechanical properties, in order to meet the needs of porous SiC ceramics in different fields[17].

碳化硅陶瓷基复合材料（CMC-SiC）是一种新型战略性热结构材料，在航空、航天、核能等高新技术领域具有广阔应用前景。但 CMC-SiC 材料硬度高、不导电等特性决定了实现其高精度、高质量加工较为困难[18]。

Silicon carbide ceramic matrix composite material（CMC-SiC） is a new strategic thermal structure material, which has broad application prospects in aviation, aerospace, nuclear energy and other high-tech fields. However, the high hardness and nonconductivity of CMC-SiC material make it difficult to achieve its high precision and high quality processing[18].

Sialon 结合 SiC 材料具有良好的抗热震性、高温强度、耐磨性、抗氧化性和抗炉渣侵蚀性等，被广泛用作高炉风口、陶瓷杯、内壁以及窑具等的材料。Sialon 结合 SiC 材料可以采用 SiC 和预合成 Sialon 烧结制得，也可以用 SiC、黏土、Al、Si 原位氮化制得。与预合成法相比，原位氮化法工艺简单，制备成本低[19]。

Sialon bonded silicon carbide materials has good thermal shock resistance, high temperature strength, wear resistance, oxidation resistance and slag erosion resistance, is widely used as blast furnace tuyere, ceramic cup, inner wall and kiln tools materials. Sialon bonded silicon carbide materials can be made by SiC and pre-synthetic Sialon sintering, or by in situ nitride with SiC, clay, Al, and Si. Compared with presynthesis, the in situ nitride process is simple and has a low preparation cost[19].

Si_3N_4/SiC 耐火材料因具有强度高、抗热震性好、抗侵蚀和耐磨性能优异等优点而广泛应用于陶瓷及高温冶金等行业。目前，在实际生产中常以纯硅粉、SiC 骨料和 SiC 细粉为原料，采用直接氮化法在 1673 K 以上温度保温较长时间（5~10 h）来制备 Si_3N_4/SiC 耐火材料；该方法存在氮化温度高、氮化周期长、氮化不完全等问题[20]。

Si_3N_4/SiC refractory material is widely used in ceramic and high-temperature metallurgy industries due to its advantages of high strength, good heat and shock resistance, erosion resistance and excellent wear resistance. At present, used pure silicon powder, SiC aggregate and SiC fine powder as the raw materials , and using direct nitride method above 1673K temperature insulation for a long time（5-10 h）to prepare Si_3N_4/SiC refractory materialsin in the actual production; This method has problems with high nitride temperature, long nitride cycle and incomplete nitride[20].

重结晶碳化硅（简称 R-SiC），是一种高 SiC 含量（质量分数在 98%以上）

的高温结构材料，它保留了 SiC 的诸多优异性能，如高温强度高、耐腐蚀性强、抗氧化性优、抗热震性好等，广泛应用于高温窑具、柴油车尾气净化器等领域[21]。

Recrystallized silicon carbide (hereinafter referred to as R-SiC), is a high SiC content (above 98% (mass fraction)) of high temperature structure material, it retains the SiC many excellent performance, such as high temperature strength, high corrosion resistance, oxidation resistance, good heat shock resistance, etc., widely used in high temperature kiln, diesel vehicle exhaust purifier, and other fields[21].

碳化硅分为两种晶型即高温型的 α-SiC 和低温型的 β-SiC。自结合碳化硅就是以低温型的 β-SiC 结合高温型的 α-SiC。其生产原理是在碳化硅颗粒中加入金属硅粉和碳（石墨、碳黑、石油焦或煤粉等）在 1450 ℃的温度下埋碳烧制使硅粉和碳反应生成低温型 β-SiC 将原碳化硅颗粒结合起来。这种结合方式的特点是利用碳化硅自身的优良特性制成性能良好的自结合碳化硅质制品。这种制品的缺点是抗氧化性较差，在陶瓷窑中用量较少[22]。

Silicon carbide is divided into two crystalline types, namely high temperature type α-SiC and low temperature type β-SiC. Self-bonded silicon carbide is the low temperature type β-SiC binding high temperature type α-SiC. The production principle is to add silicon metal powder and carbon (graphite, carbon black, petroleum coke or pulverized coal, etc.) to the silicon carbon particles burying carbon burning at 1450 ℃ reacts the silicon powder and carbon to generate low temperature type β-SiC to combine the raw silicon carbide particles. Characteristics of this combination pattern is use of the excellent characteristics of silicon carbide itself to make good self-combination of silicon carbide products. The disadvantage of this product is its poor antioxidant resistance and less dosage in ceramic kilns[22].

超高温陶瓷材料在高超声速飞行器和大气再入飞行器的应用中发挥着重要作用。然而，脆性使陶瓷材料难以适应复杂的应力应用环境。**SiC 纳米线**的引入能够提高材料的断裂韧性和抗弯强度，但 SiC 纳米线在陶瓷粉末中出现的团聚现象将显著削弱材料的力学性能。为了解决这一问题并提升材料的力学性能，Li 等采用溶胶-凝胶法在多孔 ZrB_2/ZrC/SiC 粉体孔隙中生长 SiC 纳米线，使其作为多孔超高温陶瓷材料的增强材料。Zhong 等采用聚合物前驱体裂解制备了单晶 SiC 纳米线改性的超高温陶瓷材料杂化粉体，单晶 SiC 纳米线的生长遵从 VLS 机制，可用于增强超高温陶瓷材料的力学性能[23]。

Ultra-high temperature ceramic materials play an important role in the application of hypersonic aircraft and atmospheric reentry aircraft. However, brittleness makes ceramic materials difficult to adapt to the complex stress application environment. The introduction of **SiC nanowires** can improve the fracture toughness and flexural strength of the material, but the agglomeration of SiC nanowires in ceramic powder will significantly weaken the mechanical properties of the material. In order to solve this problem and improve the mechanical properties of the material, Li et al used the sol-gel method to grow SiC nanowires in porous ZrB_2/ZrC/SiC powder pores, making them as an enhancement material for porous ultra-high temperature ceramic materials. Zhong et al used polymer precursor cracking to prepare ultra-high temperature ceramic material hybrid powder modified by single crystal SiC nanowires. The growth of single crystal SiC nanowires follows the VLS mechanism, which can be used to enhance the mechanical properties of ultra-high temperature ceramic materials[23].

碳纤维增强碳化硅陶瓷基（C/SiC）复合材料是一种新型材料，其主要成分是碳纤维和碳化硅陶瓷材料。SiC 具有优良的力学性能及抗氧化耐腐蚀的化学性能，但 SiC 断裂韧性低，脆性大。通过纤维强化制成 C/SiC 复合材料，其韧性降低、脆性减小，力学性能得到改善。与传统的结构陶瓷或碳纤维增强树脂基复合材料相比，C/SiC 复合材料的各项性能都有所提升，既具备碳纤维材料强度大、模量高、耐腐蚀、质量轻、各向异性、线膨胀系数小等特点，又兼具碳化硅陶瓷材料高抗弯性、高抗氧化性、耐腐蚀、抗磨损、摩擦系数低及高温力学性能优良等特点，还获得高抗冲击性、高抗疲劳性等优点。由于其优良的力学性能和稳定的化学性能，C/SiC 复合材料被广泛应用于能源、汽车制造、航空航天等领域[24]。

Carbon fiber reinforced silicon carbide ceramic base (C/SiC) composite material is a new type of material, its main components are carbon fiber and silicon carbide ceramic material. SiC has excellent mechanical properties and oxidation resistance, but SiC low fracture toughness, brittleness. The C/SiC composite material made by fiber reinforcement has reduced toughness, brittleness and improved mechanical properties. Compared with the traditional structure ceramics or carbon fiber reinforced resin matrix composite material, C/SiC composite material properties have improved, both carbon fiber material strength, high modulus, corrosion resistance, light quality, anisotropy, small line expansion coefficient, and silicon carbide ceramic material high bending resistance, high oxidation resistance, corrosion resistance, wear resistance, low friction coefficient and high temperature mechanical properties, also obtain high impact resistance, high fatigue resistance. Due to their excellent mechanical properties and stable chemical properties, C/SiC composites are widely used in energy, automotive manufacturing, aerospace and other fields[24].

反应烧结碳化硅具有优良的抗氧化性能，1300 ℃时氧化速率随氧化时间增加逐渐降低。随氧化时间增加，材料氧化后的常温抗折强度呈现出先增加后减小的趋势，因此对反应烧结碳化硅进行微氧化处理可以提高其常温力学性能。但氧化时间较长时，材料的力学性能会受到损害，这是因为当氧化时间较短时，碳化硅表面形成的二氧化硅具有愈合裂纹作用并有效阻碍裂纹扩展，因此材料在氧化后强度提高。随氧化时间进一步增加，碳化硅表面形成的裂纹逐渐增多，破坏了氧化膜对基体的保护，因此氧化后强度降低[25]。

The **reaction-sintered silicon carbide** has excellent antioxidant properties, and the oxidation rate gradually decreased with the increase of the oxidation time at 1300 ℃. With the increase of oxidation time, the normal temperature antibending strength of the material after oxidation shows a trend of increasing first and then decreasing, so the microoxidation treatment of the reaction sintered silicon carbide can improve its normal temperature mechanical properties. However, when the oxidation time is long, the mechanical properties of the material will be damaged. This is because when the oxidation time is short, the silica formed on the surface of silicon carbide has the effect of healing crack and effectively hindering the crack expansion, so the strength of the material is increased after oxidation. With the further increase of oxidation time, the cracks formed on the surface of silicon carbide gradually increase, which destroys the protection of the oxide membrane to the matrix, so the intensity after oxidation decreases[25].

SiC 晶须的生长一般有两种模式，分别是鳞片螺旋状和层片堆积状。鳞片螺旋状和层片堆积状 SiC 晶须，生成都包括以下步骤：（1）有机质高温裂解生成大量尺寸较小多孔碳，大量烷烃类或烯烃类小分子气体逸出；（2）添加的石英砂与多孔碳之间开始在 1200 ℃以上发生反应，伴随着 SiO 和 CO 气体生成；(3) 反应过程中，重金属氧化物被一氧化碳气体还原，形成金属液滴催化剂，促使了 SiO 进一步与多孔碳反应生成 SiC 晶须[26]。

The growth of **SiC whiskers** generally has two modes, which are scale spiral and lamellar stacking. The generation of scale spiral and layered SiC whiskers includes the following steps: (1) high temperature cracking of organic matter to produce a large amount of small porous carbon, a large number of alkanes or alkene small molecular gas escape; (2) the added quartz sand and porous carbon begin to react above 1200 ℃, with SiO and CO gas generation; (3) during the reaction process, heavy metal oxides are reduced by carbon monoxide gas to form a metal droplet catalyst, which further reacts SiO with porous carbon to form SiC whiskers[26].

碳化硅材料，特别是**固态烧结碳化硅材料（SSiC）**，在工业中广泛应用于密封、轴承和阀门的磨损部件。这是由于碳化硅材料具有良好的摩擦学性能和较高的耐腐蚀性能。**硅渗透的碳化硅（SiSiC）陶瓷**也被用于磨损部件。特别是比较大的磨损部件基本都是由 SiSiC 生产的，这是因为 SiSiC 在反应烧结过程中不发生任何收缩。SiSiC 是通过液体硅在 1500～1700 ℃温度下的渗透而产生的。由于毛细管的作用，硅发生渗透并与自由碳部分反应。由于这种反应，硅含量会降低，但 SiSiC 陶瓷通常依旧含有 5%～20%（体积分数）的游离硅，在高温和高 pH 值的溶液中不如碳化硅稳定。材料中的碳化硅相由六边形或菱面体多型（α 相）组成，它们具有相同的结构单位，但沿 *c* 轴的层顺序不同[27]。

Silicon carbide materials, and **solid state-sintered silicon carbide materials (SSiC)** in particular, have found widespread use as seals, bearings and valves in a variety of media in industrial wear applications. SiC materials are used due to their good tribological properties and high corrosion resistance. **Silicon-infiltrated SiC (SiSiC) ceramics** are also used for wear applications. Especially large wear parts are produced from SiSiC because these components can be produced without shrinkage during the reaction sintering. The materials are produced by infiltrating a SiC preform, which also contains a small amount of free carbon, with liquid silicon at a temperature of 1500 ℃ to 1700 ℃. Infiltration occurs by capillary action, with the Si partially reacting with the free carbon. Due to this reaction, the silicon content can be reduced, but SiSiC ceramics usually contain 5%-20% (volume fraction) of free silicon, which is less stable than SiC is at high temperatures and in solutions at high pH values. The SiC phase in the material consists of hexagonal or rhombohedral polytypes (α phase), which have the same structural units but differ in terms of the sequence of layers along the caxes[27].

参考文献

[1] 韩博，李进，安百俊．石墨坩埚厚度对感应加热制备太阳能级多晶硅影响的数值模拟研究［J］. 人工晶体学报，2020，49（10）：1904-1910.

[2] 王建伟，刘宁，蒋登辉，等．生物质直燃与气化对碳化硅石墨坩埚腐蚀的影响［J］. 可再生能源，2019，37（6）：809-813.

[3] 张利新，徐恩霞．石墨加入量对 Al_2O_3-SiC-C 不烧制品性能的影响［J］. 耐火材料，2017，51（4）：283-286.

[4] 徐小峰，李亚伟，王庆虎，等．硅修饰膨胀石墨对铝碳耐火材料显微结构与力学性能的

影响［J］. 人工晶体学报，2016，45（9）：2257-2264.

［5］Kim E H, Jo G H, Byeun Y K, et al. Development of a new process for the inhibition of oxidization in MgO-C refractory［J］. Journal of Ceramic Processing Research, 2013, 14（2）：265-268.

［6］Zhao Jijin, Li Xiaoxia, Guo Yuxiang, et al. Preparation and microstructure of two kinds of exfoliated graphite［J］. Advanced Materials Research, 2013, 706-708：211-214.

［7］代黎明，肖国庆，丁冬海．含碳耐火材料防氧化技术综述［J］. 材料导报，2021，35（3）：3057-3066.

［8］张红，李楠，鄢文，等．碳纤维的引入方式对铝碳耐火材料显微结构和力学性能的影响［J］. 当代化工，2022，51（7）：1527-1532，1554.

［9］陈方，叶方保，钟香崇．加入 Si，Si/Al 对高碳 Al_2O_3-C 材料抗热震性的影响［J］. 硅酸盐通报，2009，28（3）：421-425.

［10］孙旭东．连铸用含纳米碳源低碳铝碳耐火材料的研究［J］. 耐火与石灰，2019，44（5）：49-55.

［11］李韬，李亚格，吴帅兵，等．含碳耐火材料用硼基抗氧化剂研究进展［J］. 耐火材料，2023，57（2）：169-174.

［12］曹亚平．结合剂、抗氧化剂和微晶石墨对低碳镁碳砖性能的影响研究［D］. 武汉：武汉科技大学，2016.

［13］任桢，马成良，钟香崇．热处理温度对 Al 复合低碳 MgO-C 材料组成和结构的影响［J］. 耐火材料，2015，49（2）：81-85.

［14］黄军同，刘明强，孙正红，等．三辊研磨剥离鳞片石墨制备低碳含碳耐火材料用微/纳米石墨薄片［J］. 稀有金属材料与工程，2020，49（2）：682-687.

［15］Gu Qiang, Zhang Yanxiang, Ma Weikui, et al. Interfacial spinellisation of MgO-C/Al_2O_3-C composite functional refractory component at high temperatures［J］. Ceramics International, 2021, 47（2）：2705-2714.

［16］向宇，余金山，王洪磊，等．碳化硅纤维高温抗氧化性研究进展［J］. 硅酸盐通报，2022，41（9）：3234-3242.

［17］王锋，曾宇平．多孔 SiC 陶瓷制备工艺研究进展［J］. 现代技术陶瓷，2017，38（6）：412-425.

［18］王晶，成来飞，刘永胜，等．碳化硅陶瓷基复合材料加工技术研究进展［J］. 航空制造技术，2016（15）：50-56.

［19］段红娟，张向阳，韩磊，等．原位催化氮化制备 β-Sialon 结合 SiC 材料［J］. 耐火材料，2022，56（3）：189-192.

［20］韩磊，赵万国，李发亮，等．Cr_2O_3 催化氮化制备 Si_3N_4/SiC 耐火材料及其性能［J］. 机械工程材料，2018，42（10）：72-76.

［21］黄进，吴昊天，万龙刚．重结晶碳化硅材料的制备与应用研究进展［J］. 耐火材料，2017，51（1）：73-77.

［22］周丽红，王战民．碳化硅质窑具材料的结合方式及发展［J］. 耐火材料，1999（4）：234-236，239-246.

[23] 郭楚楚，成来飞，叶昉. SiC 纳米线研究进展及其应用现状［J］. 中国材料进展，2019，38（9）：831-842，886.

[24] 焦浩文，陈冰，左彬. C/SiC 复合材料的制备及加工技术研究进展［J］. 航空材料学报，2021，41（1）：19-34.

[25] 董博，余超，邓承继，等. 反应烧结碳化硅高温性能的研究进展［J］. 耐火材料，2022，56（1）：75-81.

[26] 何天颖，白建光，赵强，等. 碳化硅晶须生成机理研究［J］. 陕西科技大学学报，2022，40（1）：119-123，132.

[27] Striegler M，Matthey B，Mühle U，et al. Corrosion resistance of silicon-infiltrated siliconcarbide（SiSiC）［J］. Ceramics International，2018，44（9）：10111-10118.

5 特种耐火材料制品

5.1 氧化物制品

5.1.1 术语词组

氧化铝制品 alumina products
氧化铝基耐火陶瓷 alumina-based refractory ceramics
多孔氧化铝陶瓷 porous alumina ceramic
透明氧化铝陶瓷 transparent alumina ceramics
氧化铝坩埚 alumina crucible
氧化镁制品 magnesia products
氧化镁耐火材料制品 magnesia refractory products
氧化镁坩埚 magnesia crucible
氧化钙制品 calcium oxide products; calcia products; CaO products
氧化钙坩埚 calcia crucible; calcium oxide crucible; CaO crucible
氧化锆制品 zirconia products
高氧化锆砖 high zirconia brick
氧化锆耐火材料 zirconia refractory
氧化铬制品 chrome oxide products
高氧化铬耐火材料 high chrome oxide refractories

5.1.2 术语句子

氧化铝基耐火陶瓷由于其优异的热稳定性和耐腐蚀性能，已被广泛应用于冶金容器等高温设备中。

Alumina-based refractory ceramics have been widely used in high temperature equipment, such as metallurgical vessels, owing to their superior thermostability and corrosion resistance properties.

氧化铝坩埚已被许多研究人员普遍用于硅的溶剂提纯。

Alumina crucible has been commonly used by many researchers in the solvent refining of Si.

与手动修整相比，通过双飞秒激光扫描可以实现对这种特殊**多孔氧化铝陶瓷材料**的高质量切割，并在未来进一步应用于涡轮叶片陶瓷芯的高质量修整。

Compared to the manual trimming, a high quality cutting this special **porous alumina ceramic material** by double femtosecond laser scanning could be realized and further applied for high quality trimming of turbine blades ceramic cores in future.

使用基于挤压的3D打印机和后处理步骤（包括脱脂、真空烧结和抛光）制造**透明氧化铝陶瓷**。

Transparent alumina ceramics were fabricated using an extrusion-based 3D printer and post-processing steps including debinding, vacuum sintering, and polishing.

氧化镁制品的完整生命周期包括原料生产、背景能源生产、氧化镁产品生产、耐火材料产品生产、产品应用、废物处理和回收利用。

The complete life cycle for **magnesia products** include raw materials production, background energy production, magnesia products production, refractory products production, products application, waste products disposal and recycling.

在本工作中，使用等离子体喷涂将氧化钇涂覆在**氧化镁坩埚**上。

In the present work, plasma spraying was used to coat yttria onto **magnesia crucibles**.

氧化镁耐火材料制品是钢铁工业、水泥制造和有色冶金等高温行业窑炉的主要耐火材料。

Magnesia refractory products are the principal refractory materials for furnaces and kilns in high-temperature industries such as the iron and steel industry, cement manufacture, and non-ferrous metallurgy.

通过实验室单轴压制技术制备了**氧化钙坩埚**。

Calcia crucibles were prepared by a laboratory uniaxial pressing technique.

为了避免含CaO耐火材料在常规制备过程中的水化问题，本文采用$Ca(OH)_2$滑移铸造法制备了**CaO坩埚**。

In order to avoid the hydration problem of CaO contained refractories in the conventional preparation process, $Ca(OH)_2$ was used to prepare **CaO crucible** by slip casting in this paper.

选用高铬砖、**高氧化锆砖**、电熔 AZS 砖和 Si_3N_4 结合 SiC 砖进行耐腐蚀试验。

The high chromia brick, **high zirconia brick**, fused AZS brick and Si_3N_4 bonded SiC brick were selected for the corrosion resistance test.

此外，还将研究炉渣的熔化温度和蒸发行为，以确定**氧化锆耐火材料**随着炉渣中 CaF_2 含量的增加而出现的侵蚀现象。

Furthermore melting temperature and vaporization behavior of slag will be investigated in order to ascertain the corrosion phenomena of **zirconia refractory** with an increase of CaF_2 content in slag.

5.1.3 术语段落

辽宁菱镁矿行业的主要产品包括氧化镁、镁砖和不定形耐火材料产品。最重要的初级产品和主要原料是轻烧氧化镁、烧结氧化镁和电熔氧化镁。2014 年，该系列**氧化镁制品**的总产量为 8.7×10^6 t，占耐火材料总产量的 63%。氧化镁是使用反射窑、立窑和电炉生产的，因此在生产过程中消耗了大量的煤炭、重油和电力，导致大量的直接或间接碳排放。此外，菱镁矿是一种碳酸盐矿物，主要具有 $MgCO_3$ 化学成分，因此在矿石分解过程中会释放出一定量的 CO_2。因此，作为最重要的初级产品和生产其他耐火材料产品的主要原料，该系列氧化镁制品对菱镁矿行业的碳排放做出了重大贡献[1]。

The main products of the Liaoning magnesite industry include magnesia, magnesia brick and unshaped refractory products. The most important primary products and the main raw materials are light calcined magnesia, sintered magnesia and fused magnesia. The total output of this series of **magnesia products** in 2014 was 8.7×10^6 t, accounting for 63% of the total refractory products. Magnesia is produced using reverberatory kilns, shaft kilns and electric furnaces, and so significant quantity of coal, heavy oil and electricity are consumed during the production process, resulting in copious direct or indirect carbon emissions. In addition, magnesite is a carbonate mineral primarily having the chemical composition $MgCO_3$, and so a certain amount of CO_2 is released in the course of ore decomposition. Therefore the series of magnesia products contribute significantly to the magnesite industry carbon emissions, as the most important primary products and the main raw materials to produce other refractory products[1].

如小润湿角所示，液态 γ-TiAl 合金很好地润湿了**氧化镁、氧化钙、氧化铝和氧化钇涂层的氧化镁坩埚**。熔融的 γ-TiAl 合金严重穿透氧化镁和氧化钙坩埚，而在氧化铝和氧化钇涂层的坩埚中熔体穿透不明显[2]。

The magnesia, calcia, alumina, and yttria-coated magnesia crucibles were wetted well by the liquid γ-TiAl alloy as shown by the small wetting angle. The magnesia and calcia crucibles were severely penetrated by the molten γ-TiAl alloy, whereas melt penetration was not apparent in the alumina and yttria-coated crucibles[2].

中国是世界上最大的**镁质耐火材料及制品**生产国，导致了巨大的能源消耗和碳排放量。本研究分析了在生产阶段和使用阶段降低能耗和碳排放的措施，为可持续发展镁质耐火材料行业提供了理论依据。结果表明，**电熔镁砖**生产的**含碳镁砖**的总碳排放量远高于其他产品，**普通镁砖**和镁碳喷涂的总碳排放量均低于其他产品。**氧化镁制品**的碳排放主要集中在氧化镁的生产过程中。生产企业应选择对环境影响较小的材料，并在镁质生产过程中强调节能和减少碳排放。通过情景分析，我们发现与仅仅提高能源消耗相比，二氧化碳捕获是减少碳排放的有效措施。然而，目前还不存在一个强大的二氧化碳市场。政策制定者应计划将氧化镁行业的二氧化碳捕获纳入长期的区域二氧化碳捕获和储存发展规划。特别是，氧化镁生产集中在一个地理区域，这从而可以利用显著的规模效应。在使用阶段，延长使用寿命可以减少产品寿命内的碳排放，因此用户应尝试延长整个熔炉的使用寿命。然而，研究结果表明，增加大量的修补耐火材料来延长现有熔炉的寿命是不可取的[3]。

China is the largest producer of **magnesia refractory materials and products** in the world, resulting in significant energy consumption and carbon emissions. This study analyzes measures to reduce both the energy consumption and carbon emissions in the production phase and use phase, providing a theoretical basis for a sustainable magnesia refractory industry. Results show that the total carbon emissions of **carbon-containing magnesia bricks** produced with **fused magnesia** are much higher than those of other products, and the total carbon emissions of both **general magnesia brick** and magnesia-carbon spray are lower than those of other products. Carbon emissions of **magnesia products** are mainly concentrated in the production process of magnesia. Manufacturers should select materials with lower environmental impacts and emphasize saving energy and reducing carbon emissions in the magnesia production process. Through scenario analysis we found that CO_2 capture is an effective measure to reduce carbon emissions compared with just improving energy consumption. However, a robust CO_2 market does not currently exist. Policy makers should plan on integrating CO_2 capture in the magnesia industry into a regional CO_2 capture and storage development planning in the long term. In particular, magnesia production is concentrated in a single geographical area which would and thus could take advantage of significant scale effects. In the use phase, extending the service lifetime reduces carbon emissions over the product lifetime, thus users should attempt to extend the service lifetime of a furnace as a whole. However, results show that it is not advisable to add a large number of repair refractories to extend the lifetime of existing furnaces[3].

为避免制备过程中游离 CaO 引起的水化问题，以 $Ca(OH)_2$ 浆料为原料，采用流延法制备了**CaO 坩埚**。研究了 $Ca(OH)_2$ 浆料固体含量对 CaO 坩埚微观结构的影响。研究发现，当浆料的固体含量在一定范围内时，浆料的流变性能显著改善。浆料中固体含量的变化影响着最终产品的微观结构。当浆料的固体含量（质量分数）为 71%时，制造的 CaO 坩埚在 1600 ℃加热处理之后具有最小的平均孔径（2.58 μm）。加热的 CaO 样品具有较低的表观孔隙率（4.1%）和较高的体积密度（2.93 g/cm^3）。大多数 CaO 晶粒的尺寸分布在 10~40 μm 之间[4]。

CaO crucible was prepared from $Ca(OH)_2$ slurry via a slip-casting method in order to avoid hydration problem caused by free CaO during preparation process. The effect of solid content of $Ca(OH)_2$ slurry on microstructure of CaO crucible was investigated. It was found that the rheological property of the slurry was significantly improved when the solid content of the slurry within a certain range. The change of solid content of slurry affects the microstructure of the final products. When the solid content (mass fraction) of the slurry was 71%, the manufactured CaO crucible was of smallest mean pore size (2.58 μm) after 1600 ℃ heating treatment. The heated CaO sample was of lower apparent porosity (4.1%) and higher bulk density (2.93 g/cm^3). And the size distributions of most CaO grains are between 10 μm and 40 μm[4].

研究了四种典型耐火材料对水煤浆矿渣的抗侵蚀性能。这些典型的耐火材料分别是**高铬砖**（Cr_2O_3-Al_2O_3-ZrO_2）、熔融 AZS 砖（Al_2O_3-ZrO_2-SiO_2）、**高氧化锆砖**（ZrO_2）和 Si_3N_4 结合 SiC 砖。采用杯形试验和微观结构分析方法对其高温侵蚀性能和反应机理进行了评价。同时，通过热力学计算预测了耐火材料与矿渣之间的潜在反应。结果表明，渗透深度按 Si_3N_4 结合 SiC 砖 < 熔结 AZS 砖 < 高氧化锆砖 < 高铬砖的顺序增加。高铬砖的 ZrO_2 溶解和较高的孔隙率协同导致了最差

的耐侵蚀性。而对于高氧化锆砖，连续的炉渣渗透加速了 ZrO_2 颗粒的溶解，从而降低了其耐侵蚀性。由于孔隙率极低，可以有效地阻止炉渣渗透，熔融 AZS 砖具有更好的抗炉渣侵蚀性。对于 Si_3N_4 黏结的 SiC 砖，较差的润湿性、保护性 SiO_2 膜和增加的炉渣黏度共同有助于获得优异的耐侵蚀性[5]。

Corrosion resistance of four typical refractories against coal-water slurry coal slag, was investigated in laboratory. These typical refractories are **high chromia brick** (Cr_2O_3-Al_2O_3-ZrO_2), fused AZS brick (Al_2O_3-ZrO_2-SiO_2), **high zirconia brick** (ZrO_2) and Si_3N_4 bonded SiC brick, respectively. The cup test and micro-structural analysis method were used to evaluate the corrosion resistance and reaction mechanism at high temperature. Meanwhile, potential reactions between refractories and slag are predicted using thermodynamic calculation. The result shows that infiltrated depth increases in following order: Si_3N_4 bonded SiC brick < fused AZS brick < high zirconia brick < high chromia brick. The ZrO_2 dissolution and higher porosity for the high chromia brick synergistically result in the worst corrosion resistance. Whereas, for the high zirconia brick, the continuous slag infiltration accelerate dissolution of ZrO_2 grains, which decreases corrosion resistance. Due to the extremely low porosity, slag infiltration can be effectively hindered, and fused AZS brick has better slag corrosion resistance. For the Si_3N_4 bonded SiC brick, poor wettability, protective SiO_2 film and increased slag viscosity synergistically contribute excellent corrosion resistance[5].

氧化锆由于其优异的化学和力学性能，已成为工业上最重要的高温结构材料之一。除了高强度和韧性外，氧化锆还具有良好的硬度、耐磨性和抗热震性。氧化锆应用最广泛的领域是用于高温工艺的耐火陶瓷，如玻璃熔炉和钢连铸中的浸入式水口（SEN）。对于玻璃熔炉，用于熔炉中玻璃接触区的耐火材料必须具有对熔融玻璃的高耐侵蚀性。由于玻璃成分对耐火材料的腐蚀影响很大，因此有必要了解玻璃成分与耐火材料侵蚀之间的关系。Kato 和 Araki 根据玻璃成分研究了**含氧化锆耐火材料**的侵蚀特征。他们发现 CaO 和碱性成分的浓度对提高**氧化锆耐火材料**的侵蚀速率有很大影响[6]。

Zirconia has become one of the most industrially important high-temperature structural materials because of chemically and mechanically excellent properties. Along with high strength and toughness, zirconia also possesses good hardness, wear resistance, and thermal shock resistance. The most widely applied area of zirconia is the refractory ceramics for high-temperature process such as glass-melting furnace and a submerged entry nozzle (SEN) in continuous casting of steel. For glassmelting furnaces, the refractories used for the glass-contacting zone in furnaces must have high corrosion resistance to molten glasses. Because corrosion of refractory is greatly affected by the composition of the glasses, it is necessary to know the relation between glass compositions and refractory corrosion. Kato and Araki studied the features of corrosion of **zirconia-containing refractory** according to glass compositions. They found that the concentration of CaO as well as alkali components was very influential in increasing the corrosion rate of **zirconia refractory**[6].

气化器是用于在高温、高压和还原性气氛（低氧分压）中处理碳原料（如煤和/或石油焦）以形成 CO 和 H_2（称为合成气体）的反应容器。合成气被用作发电的燃料或化学生产的原料。气化过程的副产物包括未反应的碳、诸如 CO_2 和 H_2S 的气体，以及由碳原料中的矿物杂质或有机金属化合物形成的炉渣，这些杂质或化合物在气化过程中液化。在气化器中，炉渣与**高氧化铬耐火衬里**相互作用，通过剥落（结构和化学）和化学溶解两种主要方式造成耐火衬里的磨损并最终失效。耐火衬里的故障导致气化器停机维修，用户认为增加的使用时间对于气化作为一种工业过程的更多使用很重要。已经发现，在**高氧化铬耐火材料**中添加磷酸盐可以通过减少耐火衬里的剥落和降低化学溶解来提高商业使用期间的使用寿命。它们如何提高使用寿命的机制尚不清楚。本报告评估了商业气化炉在暴露于煤渣中约 8 个月后去除的含磷酸盐和不含磷酸盐的高氧化铬耐火材料的微观结构和物理性能，重点评估了耐火材料孔隙中矿渣/耐火材料的相互作用。介绍了研究的细节，并讨论了磷酸盐添加剂如何提高耐磨性的可能机制[7]。

Gasifiers are reaction vessels used to process carbon feedstock such as coal and/or petcoke at elevated temperature, high pressure, and in a reducing atmosphere (low oxygen partial pressure) to form CO and H_2, called synthesis gas or syngas. Syngas is used as a fuel in power generation or as a feedstock material in chemical production. By-products of the gasification process include unreacted carbon, gases such as CO_2 and H_2S, and slag formed from mineral impurities or organic metallic compounds in the carbon feedstock that liquefy during gasification. In the gasifier, slags interact with **the high chrome oxide refractory liner**, causing wear and eventual failure of the refractory lining by two primary means—spalling (structural and chemical) and chemical dissolution. Failure of the refractory lining causes the gasifier to be shut down for repair, with increased service time identified by users as important for greater usage of gasification as an industrial process. Phosphate additions to **high chrome oxide refractories** have been found to increase service life during commercial service by reducing spalling and lowering chemical dissolution of the refractory liner. The mechanism of how they improve service life is not well understood. The microstructure and physical properties of high chrome oxide refractories with and without phosphate additions removed from a commercial gasifier after approximately eight months of exposure to a coal slag are evaluated in this report, with the emphasis on evaluating slag/refractory interaction in refractory pores. Details of the investigation are presented and possible mechanisms of how phosphate additives improve wear resistance discussed[7].

5.2 烧结含锆耐火制品

5.2.1 术语词组

烧结锆莫来石耐火制品 sintered zirconia mullite refractory products

反应烧结锆莫来石复合材料 reaction-sintered zirconia-mullite composites

烧结氧化锆耐火制品 sintered zirconia refractory products

烧结锆莫来石砖 sintered zirconium mullite brick

锆刚玉砖 zircon corundum brick

烧结氧化锆陶瓷 sintered zirconia ceramics

微波烧结氧化锆 microwave sintered zirconia

5.2.2 术语句子

用作合成**烧结锆莫来石复合材料**前驱体的原材料是细粉 KF 红柱石（Imerys）、细粉锆石（Ceradel）和 CT3000SG 氧化铝（Alcoa）细粉。

The raw materials used as the precursors for the synthesis of the **sintered zirconia mullite composite** are fine powders of micronised KF andalusite (Imerys), zircon (Ceradel) and CT3000SG alumina (Alcoa).

以锆石粉和活性氧化铝为原料，加入不同比例的 MgO 和 CaO 添加剂，制备了等静压致密**反应烧结锆莫来石复合材料**。

Isostatically pressed dense **reaction-sintered zirconia-mullite composites** were prepared from zircon flour and reactive alumina with different proportions of MgO and CaO additives.

烧结氧化锆陶瓷块体成功地渗透到单体中。

Sintered zirconia ceramics blocks were successfully infiltrated with monomer.

设计师和技术人员感兴趣的两种新开发的**烧结氧化锆**的数据汇编。

Compilation of data, interesting to the designer and technologist, on two types of newly developed **sintered zirconia**.

本文主要研究了氧化钇稳定**烧结氧化锆**中韧脆转变（DBT）区的力学特征。

This work focuses on the identification of the ductile-brittle transition (DBT) zone in yttria stabilized **sintered zirconia** via the analyses of the force signatures.

5.2.3 术语段落

烧结锆莫来石和商用锆莫来石在玻璃或矿渣腐蚀开始时表现出相似的侵蚀行为。然而，在使用锆石和红柱石合成的锆莫来石复合材料的情况下，液相在高温（1200~1600 ℃之间）下的黏度仍然较高。熔融锆莫来石制品的耐侵蚀性较低：钙的渗透深度较高，液相（由 Al_2O_3、SiO_2、CaO、Na_2O、MgO、ZrO_2 等氧化物组成）在高温下（1200~1600 ℃之间）的黏性显著降低[8]。

The **sintered zirconia mullite** and the commercial zirconia mullite showed similar impregnation behaviours at the beginning of the corrosion by glass or slag. Nevertheless, the viscosity of the liquid phase at high temperature (between 1200 ℃ and 1600 ℃) remained higher in the case of the sintered zirconia mullite composite synthesized using zircon and andalusite. The fused zirconia mullite product presented a lower corrosion resistance: the depth of penetration of the calcium was higher and the liquid phase (composed of Al_2O_3, SiO_2, CaO, Na_2O, MgO, ZrO_2 oxides) was substantially less viscous at high temperature (between 1200 ℃ and 1600 ℃)[8].

为了研究 CaO 和 MgO 添加剂对**反应烧结锆莫来石复合材料**微观结构性能的一般影响，将压块在两个不同的温度（1450 ℃和 1550 ℃）下烧制，用不同量的添加剂浸泡不同的时间。对由此获得的所有压块进行 XRD 分析。为了找出在 1550 ℃下烧制的压坯中相的数量，在固定的烧制温度下，对不同量的 MgO 和 CaO 添加剂以及浸泡时间进行了广泛的 Rietveld 分析[9]。

To study the general influence of CaO and MgO additives on the microstructural properties of the **reaction-sintered zirconia-mullite composites**, the compacts were fired at two different temperatures (1450 ℃ and 1550 ℃) for different period of soaking with varying amount of additives. XRD analyses of all the compacts thus obtained were performed. To find out the quantity of phases in the compacts fired at 1550 ℃, rietveld analysis was carried out extensively with varying amount of MgO and CaO additives and soaking time for a fixed firing temperature[9].

通过在 1600 ℃下处理 2 h 的硝酸铝和锆石粉末的混合物的反应烧结制备了完全反应的锆莫来石复合材料。在通过氧化铝、氢氧化铝和锆石粉末混合物的反应煅烧制备的复合材料中检测到未反应的氧化铝颗粒。然而，通过使用氧化铝和锆石混合物可以获得相对致密的莫来石基质/氧化锆复合材料的烧结样品。在 1600 ℃下由硝酸铝和锆石制备的复合材料中，由于形成的氧化锆颗粒尺寸小且 ZrO_2 颗粒分布均匀，因此可以更好地获得四方氧化锆相。**烧结锆莫来石复合材料**的微观结构揭示了晶间和晶内 ZrO_2 颗粒的存在。晶间 ZrO_2 颗粒的形状取决于相的演变和生长条件；在 AZ 样品中，ZrO_2 颗粒是不规则的，而 AOZ 样品由于在非晶相中生长而含有球形 ZrO_2 颗粒[10]。

Fully reacted mullite-zirconia composites have been prepared by reaction sintering of mixtures of aluminium nitrate and zircon powders treated at 1600 ℃ for 2 h. Unreacted alumina particles were detected in the composites prepared by the reaction sintering of the mixtures of alumina, aluminium hydroxide and zircon powders. However, a relatively dense sintered sample of mullite matrix/zirconia composite could be achieved by the use of alumina and zircon mixtures. Tetragonal zirconia phase is better attained in composites prepared from aluminium nitrate and zircon at 1600 ℃ because of the small particle size of formed zirconia and the homogeneous distribution of ZrO_2 particles. The microstructure of **sintered zirconia-mullite composites** revealed the presence of intergranular and intragranular ZrO_2 particles. The shape of intergranular ZrO_2 particles are dependent on phase evolution and growing conditions; in the AZ sample, ZrO_2 particles are irregular, while the AOZ sample contains spherical ZrO_2 particles due to the growth in non-crystalline phase[10].

以氧化锆粉末涂层为原料，采用室温喷涂工艺制备了表面高度粗糙化的氧化锆基体。采用粉末喷涂的方法，在**烧结氧化锆**基体上沉积了均匀致密的氧化锆涂层。原子力显微镜（AFM）和表面粗糙度分析证实，氧化锆涂层的厚度和表面粗糙度随涂层循环次数的增加而增加。氧化锆涂层的四方相和化学成分与作为原料的 3Y-TZP 粉末相似，说明喷涂过程中没有发生相和成分的变化[11]。

Highly surface-roughened zirconia substrates were obtained from additive zirconia powder coating by room temperature spray processing. Homogeneous and dense zirconia coatings were deposited on **sintered zirconia** substrates with strong bonding by a powder spray coating method. The thickness and surface roughness of the coating layers on zirconia substrates increased with increasing coating cycles, which was confirmed from atomic force microscopy (AFM) and roughness analyses. The tetragonal phase and chemical composition of the zirconia coating layers were similar to those of the raw 3Y-TZP powder used as a raw material, indicating that no phase or composition changes occurred during the spray process[11].

5.3 氮化物制品

5.3.1 术语词组

反应烧结多孔氮化硅陶瓷 reaction-bonded porous silicon nitride ceramics
氮化硅基纳米复合材料 silicon nitride-based nanocomposites
氮化硅陶瓷 silicon nitride ceramics; Si_3N_4 ceramics
烧结氮化硅 sintered silicon nitride
反应结合氮化硅 reaction-bonded silicon nitride
反应烧结氮化硼 reactive sintered boron nitride
多孔氮化硼 porous boron nitride
氮化铝陶瓷 aluminium nitride ceramic

5.3.2 术语句子

为了提高氮化硅的力学性能，人们对**氮化硅基纳米复合材料**进行了多年的研究。

Silicon nitride-based nanocomposites containing SiC have been investigated over a number of years with a view to improving the mechanical properties of the Si_3N_4.

研究了氮化硅晶须对**反应烧结多孔氮化硅陶瓷**性能的影响。

The impacts of silicon nitride whiskers on the properties of **reaction-bonded porous silicon nitride ceramics** were particular studied.

这种合成方法对**氮化硼纳米材料**的结构特征、最终性能和性能有重要影响。

The synthesis approach has a significant impact on the structural features, ultimate performance and properties of **BN nanomaterials**.

我们发现**氮化硼纳米管**的力学性能敏感地依赖于诱导的力学变形水平。

We found that the mechanical properties of **boron nitride nanotubes** depended sensitively on the levels of induced mechanical deformation.

随着时间的推移，越来越多的研究人员开始设计和开展实验来研究液态锂对**氮化铝**的腐蚀行为。

As time went on, more and more researchers started to design and conduct the experiments for investigating the corrosion behaviours of the liquid lithium on **aluminium nitride**.

5.3.3 术语段落

本文综述了**氮化硅-碳化硅微纳米复合材料**的研究进展，从 Niihara 最初提出的结构陶瓷纳米复合材料的概念，到最近对这些独特材料的强度和抗蠕变性能的研究。本文描述了导致 Si_3N_4 晶粒内（晶内）和晶界（晶间）形成纳米级 SiC 的各种不同原料。后者对氮化硅的非晶界相起钉扎作用，同时也是 β-Si_3N_4 的成核位点，限制了烧结过程中晶粒的生长。这种更精细的微观结构使其强度高于单片氮化硅。由于纳米复合材料内部的残余压缩热应力，晶内 SiC 颗粒提高了强度和断裂韧性。高温强度和抗蠕变性能也比单片氮化硅高得多，本文简要回顾了这些方面的一些研究，并描述了提出的机制。在引用的其他研究的背景下，当前作者对 Si_3N_4-SiC 微纳米复合材料的研究描述了一种水处理路线，可以在烧结之前更好地分散商业粉末[12]。

This paper reviews investigations of **silicon nitride-silicon carbide micronano composites** from the original work of Niihara, who proposed the concept of structural ceramic nanocomposites, to more recent work on strength and creep resistance of these unique materials. Various different raw materials are described that lead to the formation of nanosized SiC within the Si_3N_4 grains (intragranular) and at grain boundaries (intergranular). The latter exert a pinning effect on the amorphous grain boundary phases in the silicon nitride and also act as nucleation sites for β-Si_3N_4, which limits grain growth during sintering. This finer microstructure results in strengths higher than for the monolithic silicon nitride. Intragranular SiC particles enhance strength and fracture toughness as a result of residual compressive thermal stresses within the nanocomposites. High temperature strength and creep resistance are also much higher than for monolithic silicon nitride and a few investigations of these topics are briefly reviewed and the proposed mechanisms are described. Within the context of other studies cited, work on Si_3N_4-SiC micro-nanocomposites by the current authors describes an aqueous processing route for better dispersion of commercial powders prior to sintering[12].

研究了热压氮化硅/碳化硅微/纳米复合材料以及用相同稀土氧化物烧结添加剂烧结的单片氮化硅在 70 ℃和高达 900 ℃的干滑动条件下的耐磨性。纳米复合材料的摩擦系数始终低于单片氮化硅的摩擦系数。摩擦系数随**氮化硅陶瓷**和纳米复合材料中稀土离子半径的减小而减小。同样，随着氮化硅和纳米复合材料中稀土离子半径的减小，比磨损率显著降低；掺 Lu 的陶瓷表现出最好的耐磨性。复合材料的比磨损率始终低于同等 Si_3N_4 材料。纳米复合材料的高耐磨性可以解释为更细的晶粒尺寸和更高的硬度的积极影响，克服了较低的断裂韧性的负面影响。机械磨损（微断裂）和摩擦化学反应是主要的磨损机制[13]。

The wear resistance of hot-pressed silicon nitride/silicon carbide micro/nanocomposites as well as monolithic silicon nitrides sintered with the same rare-earth oxide sintering additives was investigated under dry sliding conditions at 70 ℃ and up to 900 ℃. The friction coefficients of the nanocomposites were always lower than those of monolithic silicon nitrides. The friction coefficient decreased with decreasing ionic radius of rare-earth elements in both **the Si_3N_4 ceramics** and the nanocomposites. Similarly, the specific wear rate significantly decreased with decreasing ionic radius of rare-earths in both the silicon nitrides and the nanocomposites; ceramics doped with Lu exhibited the best wear resistance. The composites always exhibited lower specific wear rate than those of the equivalent Si_3N_4. The higher wear resistance of nanocomposites can be explained by the positive influence of finer grain size and higher hardness, which overcome the negative effects of lower fracture toughness. Mechanical wear (micro-fracture) and tribochemical reaction were found to be the main wear mechanisms[13].

加 CNF 和不加 CNF 烧结体的显微组织如图 3 所示。它们表现出类似于典型**烧结氮化硅**的棒状晶粒和玻璃晶界相。未发现碳化物相，这可能是由 CNFs 或有机黏结剂在脱黏结剂处理后残留的碳在烧结后形成的。在两种烧结体中均发现了许多孔径小于 20 μm 的孔隙。孔隙的存在使烧结体的密度相对于满密度略有下降。由于在两个烧结体中都发现了孔隙，因此 CNFs 的加入不是孔隙形成的原因[14]。

The microstructures of the sintered bodies with and without CNF are shown in Fig. 3. They exhibit rod-like grains and glassy grain boundary phases similar to those observed in typical **sintered silicon nitride**. No carbide phase, which might have formed after sintering due to carbon from the CNFs or organic binder remaining after the de-binder treatment, was found. A number of pores, with sizes less than 20 μm, were found in both sintered bodies. The existence of the pores caused a slight decrease in the density of the sintered bodies relative to the full density. Since the pores were found in both sintered bodies, the addition of CNFs was not the reason for the pore formation[14].

研究了1350 ℃和1400 ℃氮化初步合成**氮化硅**及1800 ℃和1900 ℃附加高温烧结条件对**反应结合氮化硅**性能的影响。氮化过程中形成的次生氮化硅（β-Si_3N_4）由针状颗粒组成，增强了材料的力学性能，提高了材料的机械强度。研究了氮化硅材料在不同硅和氮化硅初始配比下的微观结构和相组成。所得材料的力学性能接近液相烧结和热压氮化硅[15]。

The influence exerted by the conditions of the preliminary synthesis of **Si_3N_4** by nitridation at 1350 ℃ and 1400 ℃ and additional high-temperature sintering at 1800 ℃ and 1900 ℃ on the properties of **reaction-bonded silicon nitride** was studied. Secondary silicon nitride (β-Si_3N_4) formed in the course of nitridation consists of needle-like grains, which reinforce the material and impart to it additional mechanical strength. The microstructure and phase composition of the silicon nitride material at different initial ratios of silicon and silicon nitride were studied. The materials obtained approach liquid-phase-sintered and hot-pressed silicon nitride in the mechanical properties[15].

本文综述了类石墨六方氮化硼的晶体结构和形态特征，以及在此基础上制备材料的研究进展。介绍了热解、热压（六方和涡层改性）和**反应烧结氮化硼**的力学性能和介电性能。用有机硼和有机硅化合物浸渍氮化硼多孔样品，然后进行热解，对其物理力学特性有积极的影响[16]。

A review of the literature is presented, including a description of the crystal structure and morphological features of graphite-like hexagonal boron nitride, as well as the preparation of materials based on it. The mechanical and dielectric properties of pyrolytic, hot-pressed (hexagonal and turbostratic modification) and **reactive sintered boron nitride** are presented. The positive effect of impregnation of porous samples based on boron nitride with organoboron and organosilicon compounds, followed by pyrolysis, on the level of physicomechanical characteristics is shown[16].

多孔氮化硼作为一种新型的多孔非氧化物材料，具有比表面积大、孔径可调、化学惰性好、热稳定性好等特点，在催化、储氢、气体吸附和分离等方面具有广泛的应用前景。根据孔径的不同，将多孔氮化硼分为四种类型，即微孔氮化硼、介孔氮化硼、大孔氮化硼和层状氮化硼。本文综述了近年来不同结构多孔氮化硼的研究进展。重点介绍了不同类型的氮化硼的合成及性能，讨论了各种合成方法的优缺点。最后，展望了多孔氮化硼的应用前景和发展方向[17]。

As a new porous non-oxide material, **porous boron nitride** has attracted increasing interests due to its plenty of unique properties such as large specific surface area, tunable pore size, excellent chemical inertness and thermal stability, which are useful in catalysis, hydrogen storage, gas adsorption and separation. Based on the different pore sizes, porous boron nitride is divided into four types, namely, microporous boron nitride, mesoporous boron nitride, macroporous boron nitirde and hierarchical boron nitride. In this review, recent development of porous boron nitride with different structures is summarized. The synthesis and properties of different types of boron nitride are emphatically described, and the advantages and disadvantages of various synthesis methods are discussed. Finally, the promising applications and development directions of porous boron nitride are highlighted[17].

氮化铝（AlN）具有一系列独特的物理和化学性能，如最大的压电系数、良好的机械强度、耐腐蚀性、高击穿电压和最高的导热性。因此，铝基薄膜广泛应用于现代微电子（超声换能器、功率半导体器件、传感器等）、光电子学、光学等领域[18]。

Aluminium nitride (AlN) obtains a unique suite of physical and chemical properties, such as the largest piezoelectric coefficients, good mechanical strength, corrosion resistance, high breakdown voltage, and the highest thermal conductivity. Therefore AlN-based thin films are widely used in modern microelectronics (ultrasonic transducers, power semiconductor devices, sensors, etc.), optoelectronics, optics, and other applications[18].

氮化铝是一种111-氮化物陶瓷材料，广泛用作功能器件的组件。除了具有良好的热导率外，它在发光方面也有很高的带隙，即6 eV。在蓝宝石衬底（0001）上生长AlN薄膜。然而，两种材料之间的晶格失配导致氮化铝薄膜的微观结构存在缺陷。这些缺陷影响了氮化铝的性能。退火热处理已被以前的研究者证明是改善氮化铝薄膜微结构的最佳方法。因此，这种方法是应用在4个不同的温度，保温时间均为2 h。利用氮化铝透射电子显微镜观察退火前后金相组织的变化。我们观察到在1500 ℃时开始出现反转域。收敛束电子衍射模拟证实这些缺陷为反转域。因此，本文旨在提取制备高质量氮化铝薄膜过程中出现的问题，以及解决这一问题的方法[19]。

Aluminium nitride (AlN) is a ceramic 111-nitride material that is used widely as components in functional devices. Besides good thermal conductivity, it also has a high band gap in emitting light which is 6 eV. AlN thin film is grown on the sapphire substrate (0001). However, lattice mismatch between both materials has caused defects to exist along the microstructure of AlN thin films. The defects have affected the properties of aluminium nitride. Annealing heat treatment has been proved by the previous researcher to be the best method to improve the microstructure of aluminium nitride thin films. Hence, this method is applied at four different temperatures for two hours. The changes of aluminium nitride microstructures before and after annealing is observed using transmission electron microscope. It is observed that inversion domains start to occur at temperature of 1500 ℃. Convergent beam electron diffraction pattern simulation has confirmed the defects as inversion domain. Therefore, this paper is about to extract the matters occurred during the process of producing high quality aluminium nitride thin films and the ways to overcome this problem[19].

5.4 硼化物制品

5.4.1 术语词组

三元硼化物基陶瓷 ternary boride-based cermets
硼化物基金属陶瓷 boride-based cermets
二元硼化物陶瓷 binary boride cermets
硼化锆颗粒 zirconium boride particles
硼化锆涂层 zirconium boride coating
硼化锆-铬镍合金 zirconium boride-nichrome
硼化锆-硼化物 silicon-boron-zirconium boride

5.4.2 术语句子

B_4C、TiB_2 和 ZrB_2 三种**二元硼化物陶瓷**由于具有高硬度、高强度和优异的耐磨性而受到许多研究者的研究。

Binary boride cermets B_4C, TiB_2 and ZrB_2 have been studied by many researchers because of their high hardness, high strength and excellent wear resistance.

硼和**硼化物**在现代工业中得到了广泛的应用，本文分析了中国硼矿石的分布和特点。

Boron and **boride** have been widely applied to modern industry, and the paper analyzes the distribution and characteristic of boron ores in China.

本文分析了纳米**硼化锆**（ZrB_{12}）作为饱和吸收介质的性能。

In this article, the performance of **zirconium boride** (ZrB_{12}) nano-particles as a saturable absorber medium is analyzed.

这层保护层防止氧气渗入样品，并确保**硼化锆**不被氧化。

This protective layer prevents oxygen from penetrating into the sample and ensures against oxidation of **zirconium boride**.

在化学反应过程中，形成包裹**硼化锆**和硅颗粒的玻璃熔体，使材料具有较高的耐热性。

During the chemical reactions, a vitreous melt encapsulating the particles of **zirconium boride** and silicon is formed, which provides high heat resistance of the material.

5.4.3 术语段落

陶瓷在许多相关行业，如航空航天关键零部件、汽车、模具等，应用范围广泛，常用于制备各种耐磨、耐腐蚀的辊台、衬套、模具、工具等。**三元硼化物基陶瓷材料**由于具有高熔点、高硬度、高耐磨性和高抗氧化性等优点，被认为是最有前途的一类陶瓷材料。然而，三元硼化物基陶瓷存在脆性大、韧性低的缺陷。本文综述了三元硼化物基陶瓷的最新研究进展，重点讨论了增强剂和合金元素对三元硼化物基陶瓷力学性能的影响以及三元硼化物基陶瓷的应用。展望了三元硼化物基陶瓷的现状和研究方向[20]。

Cermets cover a wide range of applications in many related industries, such as aerospace key parts, automotive, mold, etc. , and are often used to prepare various wear-resistant and corrosion-resistant roller tables, liners, molds and tools, etc. **Ternary boride-based cermet materials** are considered to be the most promising class of cermet materials because of their advantages such as high melting point, high hardness, high wear resistance and high oxidation resistance. However, ternary boride-based cermets exhibits some defects of large brittleness and low toughness. This article summarizes the latest research progress on ternary boride based cermets, and mainly discusses the influence of mechanical properties of ternary boride based cermets and the application of ternary boride based cermets by reinforcements and alloy elements. The current situation and the research direction of ternary boride based cermets are prospected[20].

工业硼处理通常用于铁合金。由于渗硼过程是一个扩散过程，因此温度是该过程的主要参数；因此，在高温条件下也采用渗硼工艺在铁基合金表面生成**硼化物**沉积。因此，温度对达到期望的硼层厚度起着至关重要的作用。在工业上广泛使用的钢的表面存在单相（Fe_2B）或双相（Fe_2B 和 FeB）硼化物层[21]。

Industrial boron treatment is generally employed for ferrous alloys. Since the boronizing process is a diffusion-based process, the temperature is the main parameter for this process; therefore, the boronizing process is also applied at high temperatures to generate the **boride** deposits on the surface of iron-based alloys. Therefore, temperature plays a critical role to achieve the desired boron layer thickness. A single-phase (Fe_2B) or a double-phase (Fe_2B and FeB) boride layer occurs on the surface of steels that are commonly widely used in the industry[21].

本文采用放电等离子烧结法制备了两种**硼化钛复合材料**，其增强体积分数分别为 20%和 40%。利用纳米压痕技术评价了复合材料的力学性能，如断裂韧性、压痕蠕变、接触刚度、压入硬度和杨氏模量。随着硼化钛体积分数的增加，材料的弹性模量、压入硬度和接触刚度增加，而断裂韧性和压痕蠕变减小。含硼化钛体积分数为 38.5%的复合材料比含硼化钛体积分数为 24%的复合材料具有更好的力学性能。还讨论了 TiB 增强体的形态对力学性能的影响[22]。

In this paper, two **titanium boride composites** aiming at 20% and 40% (by volume) of titanium boride reinforcement were processed through spark plasma sintering. The mechanical properties such as fracture toughness, indentation creep, contact stiffness, indentation hardness and Young's modulus of the processed composites were evaluated by means of nano-indentation. The Young's modulus, indentation hardness and contact stiffness increase with the increase in volume fraction of titanium boride, while the fracture toughness and indentation creep decrease. The titanium composite with 38.5% (volume fraction) titanium boride showed improved mechanical properties compared to the composite with 24% (volume fraction) titanium boride reinforcement. The morphological influence of TiB reinforcement on the mechanical properties was also discussed[22].

研究了在真空中1300～1900 ℃范围内**硼化锆**与铬的相互作用。在固相相互作用的情况下，反应扩散发生在界面处，在金属界面处形成 Cr_2B 和 CrB。这个过程是由硼原子从二硼化锆晶格扩散到金属中所控制的。在高温范围内，相互作用通过接触熔化机制进行，是典型的共晶体系。共晶温度大约是 1780 ℃。所得到的液相以 40°的接触角湿润硼化物表面，与硼化物相互作用，并随着高温反应产物的形成而消失。它们的相组成取决于合金组分的比例，随着 ZrB_2/Cr 比例的增加而变化，代表了以硼化锆和铬晶格为基础的二元和三元化合物。考虑到准二元 ZrB_2-Cr 相图中相互作用的共晶性质，共晶在硼化锆上形成的小接触角，以及铬在气相中均匀分布的可能性，铬是一种有希望激活烧结的添加剂[23]。

The interaction of **zirconium boride** with chromium in the temperature range 1300～1900 ℃ in vacuum is studied. In case of solid-phase interaction, reaction diffusion takes place at the interface to form Cr_2B and CrB at the metal boundary. The process is controlled by the diffusion of boron atoms from the zirconium diboride lattice into the metal. In the high-temperature range, the interaction proceeds through the contact melting mechanism, being typical of eutectic systems. The eutectic temperature is about 1780 ℃. The resultant liquid phase wets the boride surface at a contact angle of 40°, interacts with the boride, and disappears since higher-temperature reaction products are formed. Their phase composition depends on the ratio of alloy components, changes with increasing ZrB_2/Cr ratio, and represents binary and ternary compounds based on the zirconium boride and chromium lattices. Given the eutectic nature of interaction in the quasibinary ZrB_2-Cr phase diagram, a small contact angle that the eutectic forms on zirconium boride, and the potential for homogeneous chromium distribution through the vapor phase, chromium is a promising addition that can activate sintering[23].

由于具有相对较低的密度、较高的熔点、较高的硬度、较强的耐腐蚀性、优异的热导率、良好的电导率等特性，ZrB_2 在结构材料中的应用越来越受到人们的关注。然而，**硼化锆**涂层的制备却难以实现。采用化学气相沉积（CVD）方法在碳化硅基体表面沉积了硼化锆。当 $ZrCl_4$ 与 $NaBH_4$ 的摩尔比为 1∶8 时，硼化锆沉积层密度最大。并对涂层的制备方法、组成及显微结构进行了研究[24]。

ZrB_2 has been attracted much attention for structural materials applications due to their unique combination of relative lower density, high melting point, high hardness, strong corrosion resistance, excellent thermal conductivity, good electrical conductivity and so on. However, the preparation of **zirconium boride** coating is difficult to realize. In this paper, zirconium boride was coated on the surface of the silicon carbide substrate by chemical vapour deposition (CVD). When the molar ratio of $ZrCl_4$ to $NaBH_4$ was 1∶8, zirconium boride deposition layer was the densest. And the preparation methods, composition and microstructure of coating were investigated[24].

5.5 硅化物制品

5.5.1 术语词组

硅化钨薄膜 tungsten silicide films
铁硅化物纳米颗粒 iron silicide nanoparticles
二硅化钼 molybdenum disilicide; $MoSi_2$
二硅化钼金属间化合物 molybdenum disilicide intermetalic

5.5.2 术语句子

研究了一种**硅化物**与锂在离子液体电解质中的反应行为。

We investigated the reaction behavior of a **silicide** with Li in an ionic-liquid electrolyte.

为了提高**硅化镍**的稳定性，铂合金化镍膜得到了广泛的应用。

Nickel film alloyed with platinum was widely used to improve the stability of **nickel silicide**.

机械合金化是一种高强度的高能铣削工艺，已被用于从元素粉末共混物中合成**二硅化钼**。

Mechanical alloying is an intensive high-energy milling process and has been used to synthesize **molybdenum disilicide** from elemental powder blends.

在**二硅化钼**对应的组合物中，Mo-Si 混合物的圆柱形压块在从点火线圈接收到足够的辐射能后容易被压块的表层点燃。

Cylindrical compacts of Mo-Si mixture in the composition corresponding to **molybdenum disilicide** ignited easily after a sufficient amount of radiant energy from the ignition coil was received by the surface layer of the compact.

$MoSi_2$（**二硅化钼**）由于其低密度（6.31 g/cm^3）、高熔点、高导电性以及在高温下（甚至在腐蚀性极强的环境下）非常好的抗氧化性而引起了人们极大的研究兴趣。

$MoSi_2$（**molybdenum disilicide**）has attracted great research interest due to its rather low density（6.31 g/cm^3），high melting point，high electrical conductivity and very good oxidation resistance at high temperature，even in very aggressive environments.

5.5.3 术语段落

硅化物是一种有吸引力的新型活性材料，可用于使用某些离子液体电解质的下一代锂离子电池的负极；然而，上述组合的反应机理尚不清楚。在硅化物电极上可能发生的反应有：锂金属在电极上的沉积和溶解，硅的锂化和去硫化，这是由硅化物的相分离引起的，硅与锂的合金化和去硫化。在这里，我们使用各种分析方法检查了这些可能性。结果表明，硅化物发生了锂化和去硫化[25]。

Silicides are attractive novel active materials for use in the negative-electrodes of next-generation lithium-ion batteries that use certain ionic-liquid electrolytes；however，the reaction mechanism of the above combination is yet to be clarified. Possible reactions at the silicide electrode are as follows：deposition and dissolution of Li metal on the electrode，lithiation and delithiation of Si，which would result from the phase separation of the silicide，and alloying and dealloying of the silicide with Li. Herein，we examined these possibilities using various analysis methods. The results revealed that the lithiation and delithiation of silicide occurred[25].

这项工作证明了优化**硅化镍（NiSi）**纳米线（NWs）生长的重要因素。在不同的沉积温度下制备了样品并对其进行了表征。研究了金属与硅的自发反应形成硅化物。金属扩散是形成硅化物相和层状形貌的关键因素。在特定条件下沉积的初始 NiSi 层比初始钴（Co）硅化物层诱导的 NiSiNW 生长更快。在单硅化镍模板上制备 NWs，Ni 是比 Si 更快的扩散剂。初始阶段显著影响硅化物形成的晶粒尺寸（180~368 nm）。通过后退火工艺研究了 NWs 的热稳定性。通过对加热过程的调制进行了系统的论证。我们报道了压缩应力是 NWs 生长的驱动力[26]。

This work demonstrates the important factors to optimize the growth of **nickel silicide**（NiSi）nanowires（NWs）. The samples were prepared at different deposition temperatures and characterized. Spontaneous reaction of metal and Si has been investigated for silicide formations. Metal diffusion is a crucial factor to form silicide phases and layer morphologies. The initial NiSi layer deposited under specific conditions induced fast NiSiNW growth rather than initial cobalt (Co) silicide layer. Ni is a faster diffuser rather than Si to produce NWs on Ni monosilicide template. The initial stage significantly affects silicide-formed grain sizes (180-368 nm). Thermal stability of the NWs was studied via postannealing process. Systematic demonstration has been performed by modulation of heating processes. We report that compressive stress is a driving force for the growth of NWs[26].

反应蒸汽渗透（RVI）也被用于生产**二硅化钼（$MoSi_2$）**。以多孔粉末致密体形式存在的钼被硅蒸气转化为 $MoSi_2$。在 RVI 工艺中，松散压实的金属钼粉在略低于硅熔点的温度下暴露于硅蒸气中。硅蒸气可以通过 $SiCl_4$ 在氢气存在下的气相分解提供。样品的横截面显示出较厚的 $MoSi_2$ 外层、较薄的 Mo_5Si_3 中间层和未反应的钼芯。这些观察结果表明，在该过程开始时，钼与硅反应形成 Mo_5Si_3，硅的进一步供应将 Mo_5Si_3 转化为 $MoSi_2$。随着这一过程的进行，硅扩散穿过表面的 $MoSi_2$ 层，依次将每个钼颗粒转化为 Mo_5Si_3，然后转化为 $MoSi_2$，这样 Mo_5Si_3 和 $MoSi_2$ 的不同锋面逐渐从致密体表面向内移动[27]。

Reactive vapor infiltration (RVI) has also been explored to produce **molybdenum disilicide ($MoSi_2$)**. Molybdenum in the form of a fine porous powder compact is converted to $MoSi_2$ by silicon vapor. In the RVI process, a loosely compacted metallic molybdenum powder is exposed to silicon vapor at temperatures slightly below the melting point of silicon. Silicon vapor can be supplied through a gas-phase decomposition of $SiCl_4$ in the presence of hydrogen gas. The sample cross sections show a thick $MoSi_2$ external layer, a thin Mo_5Si_3 intermediate layer, and an unreacted molybdenum core. These observations suggest that, at the onset of the process, molybdenum reacts with silicon to form Mo_5Si_3 and a further supply of silicon converts Mo_5Si_3 to $MoSi_2$. As the process proceeds, silicon diffuses through the surface $MoSi_2$ layer, successively converting each molybdenum particle to Mo_5Si_3 and then to $MoSi_2$ so that distinct fronts of Mo_5Si_3 and $MoSi_2$ move progressively inward from the surface of the compact[27].

硅与钼的原子比为 1.0 时，硅是形成二硅化钼的限制反应物，因此硅在反应物中的扩散是非常重要的。尽管硅在钼和二硅化钼中的扩散活化能分别为 183.9 kJ/mol 和 240.8 kJ/mol，但在低温条件下，单质钼和硅之间仍可能发生固态反应。Angelescu 报道，**二硅化钼（$MoSi_2$）**的形成始于 1100~1300 ℃的温度范围。Deevi 根据 Mo+2Si 混合物的 X 射线图谱，证明钼和硅之间的固体扩散反应发生在 1200 ℃[28]。

Given that silicon is the limiting reactant in the formation of molybdenum disilicide at the atomic ratio of silicon to molybdenum of 1.0, the diffusion of silicon in the reactants is very important. Even though the respective activation energies for silicon diffusion in molybdenum and molybdenum disilicide is 183.9 kJ/mol and 240.8 kJ/mol, the solid-state reaction between elemental molybdenum and silicon may occur at low temperatures. Angelescu reported that the formation of **molybdenum disilicide ($MoSi_2$)** was initiated at temperatures in the range of 1100-1300 ℃. Deevi insisted that the onset of the solid diffusional reaction between molybdenum and silicon occurred at the lower temperature of 1200 ℃ based on the X-ray phases of the Mo + 2Si mixtures[28].

采用“化学炉”燃烧合成法，添加 NH_4Cl 助剂合成了高纯单相**二硅化钼金属间化合物**。采用硅钛混合粉作为化学烘箱，提高了 SHS 工艺对燃烧产物的净化效果。在反应物中加入 NH_4Cl 以控制颗粒生长和均质产物。含 NH_4Cl 的反应物不能在正常的 SHS 中被点燃。结果表明，这种燃烧合成方法对于制备一些传统燃烧合成法无法得到的粉末是一种潜在的有用方法[29]。

High pure single-phase **molybdenum disilicide intermetalic** was synthesized by " chemical oven" combustion synthesis method with adding NH_4Cl additive. The mixed powders of Si and Ti were used as chemical oven to enhance SHS procedure for purifying combustion products. NH_4Cl was added to reactants for controlling grains growth and homogenizing products. The reactants with NH_4Cl cannot be ignited in normal SHS. It can be concluded that this process of combustion synthesis is a potential useful method for preparing some powders that cannot be obtained through conventional SHS[29].

5.6 功能耐火材料制品

5.6.1 术语词组

滑动水口 sliding nozzle; sliding gate
浸入式水口 submerged entry nozzles
旋涡浸入式水口 the submerged entry nozzles with swirling flow
长水口 long nozzles
整体塞棒 monolithic stopper
透气元件 porous plugs
定径水口 metering nozzles

5.6.2 术语句子

填充砂通常用于钢包的滑动水口，以使**滑动水口**耐火材料与钢水分开。

Filler sands are usually used for the **sliding nozzle** of a ladle to keep the sliding gate refractory apart from molten steel.

为了降低样品中 SiO_2 的浓度，提高其高温性能和抗渣性，用 m-ZrO_2 或 ZA 代替 ZAS 作为 ZrO_2 化合物的选择，利用回收耐火材料制备 **Al_2O_3-ZrO_2-C 滑动水口**。

In order to decrease the concentration of SiO_2 in the sample, as well as improve its high temperature performance and slag resistance, the ZAS is substituted by m-ZrO_2 or ZA as the selection of the ZrO_2 compounds for preparing the **Al_2O_3-ZrO_2-C sliding nozzle** using recovery refractory.

炼钢厂连铸机旋涡**浸入式水口**的开发，始于对多股圆坯连铸机旋涡叶片的耐久性和旋涡流强度的研究，因为圆坯连铸机对铸造作业和产品质量的风险相对较小。

The development of **submerged entry nozzles** with swirling flow for continuous casters in steelmaking works started with investigation of the durability of the swirl blade and intensity of swirling flow in the multi-strand round billet caster since the risk to casting operation and the quality of products for round billets was considered relatively small.

浸入式水口（SEN）的主要用途是将钢液从中间包输送到模具中。

The main purpose of a **submerged entry nozzle（SEN）** is to distribute molten steel from a tundish to a mold.

采用低碳**长水口**、中间包低碳涂层、低碳模具粉末等措施防止吸碳。

Measures adopted for prevention of carbon pick-up are **long nozzles** with less carbon, coating material with low carbon content for the tundish, and low carbon mould powder.

在连铸过程中，为了防止钢水中的活性元素氧化，钢包与中间包之间使用了铝碳**长水口**。

The **long nozzles** made of alumina-carbon both for the steel ladle and the steel tundish were used in the process of continuous casting to prevent the active elements in the molten steel from oxidation.

连铸是钢铁生产中的一项重要工艺，用于连铸的耐火材料主要有**长水口**、**整体塞棒**、**浸入式水口**等功能耐火材料。

Continuous casting is an important process in steel production and the refractories used for continuous casting mainly contain **long nozzle**, **monolithic stopper**, **submerged nozzle** and other functional refractories.

尖晶石-C 对于**整体塞棒**和**浸没式水口**很重要，通常决定了应用效果。

Spinel-C material is important for **monolithic stopper** and **submerged nozzle**, which usually determine the application effects.

从以上三相可以看出，当使用氧化铝碳**整体塞棒**时，Al_2O_3 与 FeO 或 MnO 在铸造温度下反应很容易形成液相。

From the above three phases we can see that when the alumina carbon **monolithic stopper** is used, the liquid phase is easily formed by reaction of Al_2O_3 with FeO or MnO at the temperature of casting.

本工作的目的是建立一个数学模型，以研究**透气元件**的数量及其径向位置对热和化学混合现象的影响，考虑了工业钢包中三个电极的局部加热源。

The objective of this work was to develop a mathematical model to investigate the influence of the number of **porous plugs** and its radial position on mixing phenomena both thermal and chemical, considering the localized source of heating from three electrodes in an industrial ladle.

过去，人们研究了钢包内**透气元件**的数量、位置和相对角度以及气体流量等因素对混合时间的影响，并在工业实践中起到了显著的作用。

In the past, the effects of many factors such as the number and position and relative angle of **porous plugs** in a ladle, as well as gas flowrate on the mixing time have been studied and some beneficial results used in industrial practice.

定径水口和油润滑的连铸钢坯是获得民用建筑长产品的有效工具，具有高生产率、低成本和较高的质量。

Continuous casting of billets with **metering nozzle** and oil lubrication is an efficient tool for obtaining long products for civil construction, with high productivity, low cost, and a quality level acceptable for these applications.

显然，该技术的一个好处是通过浸入式水口和整体塞棒或其他流量计量装置取代通过**定径水口**的开流浇注，从而减少了模具湍流。

Obviously one of the benefits of this technology is the reduction in mould turbulence by the replacement of open-stream pouring through **metering nozzles** with submerged entry nozzles and stopper rods or some other flow metering device.

5.6.3 术语段落

计算结果与工业观测结果吻合较好：**滑动水口**内孔堵塞强度随中心滑动闸板对铸造通道的封闭程度而增大。这种行为应与非金属夹杂物对钢的污染、铸钢温度以及滑动水口和中间包水口耐火材料的导热系数等因素的作用增加有关。这种情况会显著缩短滑动水口的使用寿命[30]。

The calculation results agree well with the industrial observation results: the hole clogging intensity in the **sliding gate** increases with the degree of closure of the casting channel by the central sliding gate plate. This behavior should be related to an increase in the role of factors such as the contamination of steel by nonmetallic inclusions, the casting steel temperature, and the thermal conductivity of the refractory materials of sliding gate plates and tundish nozzle. These circumstances can significantly decrease the sliding gate lifetime[30].

在和 Wakayama 和 Kashima 工厂的全尺寸水模试验和工业铸造试验中，设计并评估了板坯连铸用**旋涡浸入式水口**。这些实验清楚地表明，适当强度的浸入式水口内的旋流产生与适当的出口设计相结合，显著提高了模具内流动的稳定性，减少了镀锌汽车板用超低碳钢板坯和钢板的表面缺陷，超低碳钢是性能改进最明显的钢种之一。在前人研究的基础上，设计了旋涡式浸入式水口。采用带旋流的浸入式水口的效果与在模具中使用电磁流量控制装置的效果相当[31]。

The submerged entry nozzles with swirling flow for slab casting were designed and evaluated in full-scale water model experiments and industrial casting examinations at the Wakayama and Kashima works. These experiments clearly showed that the combination of the swirling flow generation with proper intensity in the submerged entry nozzle and adequate design of the outlet ports improves the stability of flow in the mold remarkably, and reduces surface defects on slabs and steel sheets of ultra low-carbon steel destined to be galvanized automobile panels, which are one of the most severe grades of steel. The submerged entry nozzles with swirling flow were designed on the basis of previous research. The effect using the submerged entry nozzles with swirling flow was comparable to that of using electromagnetic flow control devices in a mold[31].

在**浸入式水口（SEM）**中，钢水通过圆柱形管道，内壁通常由氧化铝制成。钢和水口耐火材料之间的反应可能导致堵塞。中间包的重力和金属总重量将流体沿水口输送；SEN 下区有两个外侧出口。钢液流与浸没式水口底部和内壁相互作用，形成两个旋涡[32]。

In a **submerged entry nozzle（SEM）**, the molten steel passes through a cylinder-shaped pipe, internal walls are usually made of alumina. The reaction between the steel and the nozzle refractory could cause clogging. The gravitational force and the total metal weight from tundish send the fluid along the nozzle; there are two lateral exit ports in the SEN lower region. The liquid steel flow interacts with the bottom and the inner walls of the submerged nozzle, and develops two vortexes[32].

在钢水从钢包转移到中间包的过程中，**长水口**和包覆管（包括其变体）是保护钢水不被空气二次氧化的两种常用手段。每种方法都有其优点和缺点。长水口由于其方便的自动安装而得到了更广泛的应用。它们占用空间小，并且与中间包结构隔离。但是，长水口需要在加热后更换。如果钢包打开失败，则需要拆除长水口，对钢包水口进行氧清洗。水口的水下开口清洗时需要小心，以尽量减少剧烈的反冲。在惰性气氛下沿着水口打开钢包进行第一次加热是非常困难的[33]。

The **long nozzle** and the shrouding pipe（including their variants）have been the two common means to protect steel melt from reoxidation by air during melt transfer from the ladle to the tundish. Each of these methods has its advantages and disadvantages. Long nozzles have found more popular use because of their convenient automatic mounting. Their small space occupancy, and their isolation from the tundish structure. However, the long nozzle needs to be replaced heat after heat. In case the ladle opening fails, the long nozzle needs to be removed for oxygen cleaning of the ladle nozzle. The submerged opening of the nozzle requires care to minimize aggressive blow backs. Ladle opening with along nozzle under an inert atmosphere for the first heat may be more difficult[33].

这些改进经受住了时间的考验，在成本、生产力和质量方面证明了它们的可行性。它们在处理稳态铸造方面已经接近成熟。事实上，一个充满惰性气氛的密封中间包，其容量相当大，具有可接受的宽深比，熔池深度超过 1 m，实践证明其能够在没有任何额外污染的情况下将干净的熔体输送到模具中。熔体通过浸入式**长水口**或包覆管从钢包浇注到中间包。这些措施与带有惰性气体预热和等离子体或感应中间包加热器的热中间包相结合，提高了连铸机的生产率，而不会对钢的质量造成太大损害[34]。

These improvements have survived the test of time, confining their viability in terms of cost, productivity, and quality, they have come to near maturity in dealing with the steady state casting. In fact, a sealed tundish filled with inert atmosphere and of a reasonably large capacity of acceptable width to depth ratio with a bath depth of more than 1 m has proven capable of delivering clean melt to the mold without any additional contamination. The melt is poured from the ladle to tundish through an immersed **long nozzle** or a shrounding pipe. These measures together with a hot tundish cycle with inert gas preheating and plasma or induction tundish heaters have increased the productivity of continuous casters without much impairment of steel quality[34].

含碳耐火材料因其优异的抗热震性和耐腐蚀性成为制备**长水口**、**整体塞棒**和**浸入式水口**的主要材料。以往的研究主要集中在用于连铸的低碳镁碳耐火材料、铝碳耐火材料和尖晶石碳材料。Yong Cheng 等研究了片状石墨含量对镁碳耐火材料显微组织和性能的影响。结果表明：碳含量的增加会降低试样的断裂强度和热膨胀率，而镁碳耐火材料的导热性和抗热震性增强[35]。

Carbon containing refractories become the main materials for preparing **long nozzle**, **monolithic stopper** and **submerged nozzle** because of the outstanding thermal shock resistance and corrosion resistance. Previous studies mainly focus on the low carbon magnesia-carbon refractories, alumina-carbon refractories and spinel-carbon materials used for continuous casting. Yong Cheng et al studied the effect of flake graphite content on the microstructure and properties of MgO-C refractories. The results showed that increasing carbon content could reduce the fracture strength and thermal expansion rate of the specimens, whereas the thermal conductivity and thermal shock resistance of MgO-C refractories enhanced[35].

尖晶石-碳材料通常用于**整体塞棒**和**浸入式水口**的材料，这些材料对抗热震性能有苛刻的要求。材料的组成决定了材料的抗热震性，但很少有论文系统地研究尖晶石-碳材料的抗热震性。本文研究了添加剂、不同种类石墨和沥青对尖晶石-碳材料抗热震性能的影响[36]。

Spinel-C material is usually used in the head of **monolithic stopper** and the bowl of **submerged nozzle**, which require excellent thermal shock resistance. The composition determines the thermal shock resistance of materials, but few papers systematically study of the thermal shock resistance of spinel-C material. In this paper, the effects of additives, different kinds of graphite and pitches on thermal shock resistance of spinel-C material were investigated[36].

材料 B 的铸造时间比 A 长，B 的热值是 A 的 3~4 倍，但离目标还有一定的距离。材料 C 是改进的尖晶石-碳材料，其中形成陶瓷和碳键的复合体，并原位形成 AlN。热值为 A 的 4~6 倍，用材料 C 制成的**整体式塞棒**基本可以满足钢厂的需求[37]。

The casting time for materials B is longer than that of A and the heat number is 3 to 4 times, but there is still some distance from the target. Materials C are improved spinel-carbon materials, in which the composite combination of ceramic and carbon bond is formed and AlN is in situ formed. The heat number is 4 to 6 times, and the **monolithic stopper** made of material C can basically meet the needs of the steel plant[37].

因此，确定合适的氩气流量具有重要意义。此外，在国内许多钢厂使用的大容量钢包中，通常使用两个**透气元件**，它们的位置和相对角度对钢液的均匀化有重要影响[38]。

Thus, it is of great significance to determine proper argon stirring flowrate. Moreover, for high-capacity ladles used in many steel factories of China, two **porous plugs** are usually utilized, and their positions and relative angle have effect on the homogenization of molten steel[38].

透气元件的位置相当靠近钢包壁，这导致钢包上部的羽流和钢包壁之间有相对强烈的再循环回路。这可能会导致耐火材料磨损。羽流向钢包中心弯曲。这是因为钢包壁靠近羽流会将钢材“推”到钢包中心[39]。

The **porous plugs** are located rather close to the ladle walls, which causes relatively strong recirculation loops between the plumes and ladle walls in the upper part of the ladle. This might cause wear of the refractory material. The plumes are bent towards the ladle centre. This is because the nearness of the ladle wall to the plume "pushes" the steel against the ladle centre[39].

硅、锰和少量铝镇静的钢通过**定径水口**进行铸造。通常在低碳钢中可以有意加入金属铝或者合金（进行镇定）。这些元素发生氧化速率的顺序为铝、硅和锰。(由于发生氧化）导致形成了较微米尺寸夹杂物的，富含氧化硅和氧化锰，但氧化铝含量较低的再氧化夹杂物，这一现象已由 Farrell 和 Hilty 描述报道[40]。

Steels for casting with **metering nozzle** are usually killed with silicon, manganese and a small amount of aluminum. Aluminum may be added purposely, as is normal in low carbon steel, or via ferroalloys. The elements that will oxidize faster are aluminum, silicon and manganese. The result is that reoxidation inclusions are formed, rich in silica and manganese oxide, but with lower alumina content in comparison with microinclusiones, as early described by Farrell and Hilty[40].

在开流浇注中，钢液通过**定径水口**从中间包底部流出，流穿过空气或气体罩落入模腔。中间包流的粗糙度已被证明对空气夹带有较大的影响。夹带的气泡在模腔中上升，并在弯月面处产生明显的扰动。在 20 世纪 70 年代初由 McLean 和同事进行的一项特殊研究中，报告了中间包流质量的可变性及其对弯月面处湍流和钢坯质量的影响[41]。

In open stream pouring, liquid steel issues from the bottom of the tundish via a **metering nozzle** and the stream falls through air or a gas shroud into the mould cavity. The roughness of the tundish stream has been shown to have a profound effect on air entrainment. The entrained gas bubbles rise up in the mould cavity and create significant disturbances at the meniscus. In an exceptional study conducted by McLean and co-workers in the early seventies, the variability of tundish stream quality and its effects on meniscus turbulence and billet quality have been reported[41].

5.7 节能耐火材料制品

5.7.1 术语词组

隔热耐火材料 refractory thermal insulation materials; insulation refractories

轻质隔热耐火材料 lightweight insulation refractories

高隔热耐火材料 high-refractory thermal insulating materials

多孔隔热耐火材料 porous refractory thermal insulating materials

不烧隔热耐火材料 unroasted thermal insulating refractory materials

膨胀蛭石 expanded vermiculite

硅藻土 diatomite
天然硅藻土 natural diatomite
硅藻土复合水凝胶 the diatomite composite hydrogel
粉煤灰 fly ash
硅质粉煤灰 siliceous fly ash
钙质粉煤灰 calcareous fly ash
粉煤灰悬浮液 fly ash suspension
漂珠 floating bead
漂珠陶瓷 floating bead ceramics
粉煤灰漂珠 fly ash floating beads
耐火氧化物空心球 hollow spheres of refractory
氧化铝空心球 alumina hollow spheres
$TiZrO_4$ 空心球 $TiZrO_4$ hollow spheres
膨胀珍珠岩 expanded perlite
石蜡/膨胀珍珠岩材料 paraffin/expanded perlite material
轻质黏土砖 lightweight clay bricks
不烧轻质黏土砖 unfired lightweight clay bricks
多孔轻质黏土砖 porous lightweight clay bricks
环保轻质黏土砖 environmentally friendly lightweight clay bricks
莫来石系轻质砖 lightweight mullite bricks
轻质莫来石基莫来石-碳化硅砖 lightweight mullite-based mullite-SiC bricks
轻质高铝砖 lightweight high alumina bricks
轻质砖 lightweight bricks
氧化铝轻质保温砖 alumina lightweight insulation brick
免烧高铝砖 unfired high alumina bricks
轻质硅砖 lightweight silica bricks
钙长石轻质隔热砖 lightweight anorthite insulation bricks
钙长石基隔热耐火砖 anorthite based insulating firebricks
石棉 asbestos
温石棉 chrysotile
石棉布复合纳米材料 asbestos cloth nanocomposite
石棉酚醛复合材料 asbestos-phenolic composite
玻璃棉 glass wool
岩棉 rock wool
硅酸铝耐火纤维 aluminium silicate refractory fibre
多晶氧化铝纤维 polycrystalline alumina fiber
多晶氧化铝基纤维 polycrystalline alumina-based fibers
氧化锆纤维 zirconia fiber
纤维状多孔氧化锆陶瓷 fibrous porous zirconia ceramics
超细二氧化硅微粉 ultrafine silica powders
多孔硅酸钙 porous calcium silicate
微孔硅酸钙 microporous calcium silicate
硅酸钙基材料 calcium silicate-based materials
浇钢用隔热板 insulation boards for steel casting
气凝胶超级隔热材料 aerogel super insulation materials
亚晶锆石纳米纤维气凝胶 hypocrystalline zircon nanofibrous aerogels
陶瓷气凝胶 ceramic aerogels
微孔轻质隔热材料 microporous lightweight insulator materials
微孔耐火材料 microporous refractories
轻质微孔镁基耐火材料 lightweight microporous magnesia-based refractories
轻质 Al_2O_3-MgO-C 耐火材料 lightweight Al_2O_3-MgO-C refractories
轻质氧化镁 lightweight magnesia
节能涂料 energy-saving coatings
高温红外辐射节能涂层 high temperature infrared radiation energy saving coating
远红外涂料 far-infrared coating
高发射率涂层 high emissivity coatings

5.7.2 术语句子

以 626 K、773 K 和 910 K 附近的剥离蛭石为例，利用加法实验数据对**多孔隔热耐火材料**的三种传热机理进行了估算。

The results of estimation of three heat transfer mechanisms in **porous refractory thermal insulating materials** on example of exfoliated vermiculite at temperatures near 626 K, 773 K and 910 K using experimental data by additive approach are presented.

这一特点表明，这些废弃材料有可能成为制备**隔热耐火材料**的潜在材料。

This promising characteristic suggests that these wastes materials may lead to be used as a potential material for the preparation of **insulation refractories.**

轻质隔热耐火材料对高温性能至关重要，以降低能源消耗。

Lightweight insulation refractories are essential for high-temperature performance to reduce energy consumption.

结果表明，与疏水蛭石相比，**膨胀蛭石**对油的亲和力更强。

The results showed that the **expanded vermiculite** had a greater affinity for oil than hydrophobized vermiculite.

当**膨胀蛭石**以 30% 和 60% 的替代水平加入时，与没有膨胀蛭石的普通砂浆相比，流动直径更大。

When the **expanded vermiculite** was included at 30% and 60% replacement levels, the flow diameter was higher compared to the plain mortar without expanded vermiculite.

因此，**硅藻土**可以改善 PAAm（聚丙烯酰胺）水凝胶的力学性能，延长 PAAm 水凝胶负载多菌灵的药效。

Thus, **diatomite** can improve the mechanical property of PAAm hydrogel and extend the carbendazim efficacy loaded with PAAm hydrogel.

考虑到更广泛的适用性，**硅藻土复合水凝胶**可以应用于减少药物的损失和浪费，延长药物在生物医学材料中的使用期限。

Considering more widely applicable, **the diatomite composite hydrogel** can be applied to reduce the loss and waste of drugs and prolong the duration of drugs in biomedical material.

粉煤灰由结晶相和非晶相组成。

Fly ash consists of crystalline phase and amorphous phase.

作为固体废料的**粉煤灰**可用作废水处理的吸附剂。

Fly ash which is solid waste can be used as an adsorbent for waste water treatment.

使用粒径为0.12~0.18 mm的**漂珠**可以烧结具有高孔隙率、低体积密度和适当抗弯曲强度的陶瓷体。

A ceramic body with high porosity, low volume density, and appropriate resistance to bending strength can be sintered by using **floating bead** with particle size of 0.12-0.18 mm.

根据具体的要求，**漂珠陶瓷**的性能可以通过改变**漂珠**的粒径来提高。

The performance of the **floating bead ceramics** can be improved by changing the particle size of the **floating bead** according to specific requirements.

当**粉煤灰漂珠**的含量为5%时，可使耐火涂料的耐火性能达到最佳。

When the content of **fly ash floating beads** was 5%, it could make the fire-resistant properties of fire-resistant coatings to achieve the best.

氧化铝是一种常见的工业材料，可在1400~2000 ℃高温炉中熔炼制备**氧化铝空心球**。

Alumina is a common industrial material, which can be smelted in 1400-2000 ℃ high-temperature furnaces to prepare **alumina hollow spheres**.

在这些结构中，**$TiZrO_4$ 空心球**具有密度低、表面渗透性好、电子转移效率高等特点，最终被用于我们的工作中。

Among these structures, **$TiZrO_4$ hollow spheres** have low density, good surface permeability, and high efficiency of electron transfer, and they were finally used in our work.

在此，我们设计了一种简便且经济高效的**空心球形 Bi_2WO_6/还原氧化石墨**

烯（RGO）复合材料的可控合成路线，该复合材料可有效降解五种有机废水，包括罗丹明 B、甲基橙、苯酚、磺胺间甲氧嘧啶和磺胺。

Herein, we design a facile and cost-effective route for the controllable synthesis of **hollow sphere shaped Bi_2WO_6/reduced graphene oxide (RGO) composites**, which can efficiently degrade five organic waste water, including rhodamine B, methyl orange, phenol, sulfamonomethoxine and sulfanilamide.

膨胀珍珠岩作为支撑材料，不仅为整个相变材料提供了良好的机械强度，而且还防止了相变材料的泄漏。

Expanded perlite as supporting material not only provides a good mechanical strength to the whole phase change material (PCM), but also prevents the leakage of PCM.

得出的结论是，**轻质黏土砖**是具有改善热性能的良好绝缘材料。

Concluding that **lightweight clay bricks** are good insulating materials with improved thermal properties.

本文讨论了将再生塑料废料作为建筑材料添加剂加以充分利用的潜力，以生产**不烧轻质黏土砖**，并重点关注其热性能。

This paper discusses the potential of putting the recycled plastics waste into good use, as a construction material additive, to produce **unfired lightweight clay bricks** focusing on their thermal performance.

与含有原始轻质骨料的试样相比，硅溶胶涂层的**轻质莫来石基莫来石-碳化硅砖**可以解决轻质材料的问题，可以抵抗碱侵蚀，并具有更高的抗压强度。

Silica sol-coated **lightweight mullite-based mullite-SiC bricks** could solve the problem of lightweight materials that resist alkali attack and lead to higher compressive strength compared with specimens that contain pristine lightweight aggregates.

同时，气孔的大小、形状和分布使**氧化铝轻质保温砖**具有较强的抗压强度和较低的热导系数。

At the same time, the size, shape and distribution of pore make **the alumina lightweight insulation brick** has strong compressive strength and low thermal conductivity.

轻质砖是一种建筑用砖，由硅砂制成，具有耐热性，可降低火灾风险。

Lightweight brick is a kind of brick for building was made from silica sand which is resistant to heat and can reduce risk fire.

为了找出**轻质砖**的特征行为的程度，使用水泥、沙子、泡沫剂和硅灰等材料进行了抗压强度和吸收性研究。

To find out the extent of the behavior of the characteristics of **lightweight bricks**, a study of compressive strength and absorption was carried out using materials such as cement, sand, foam agents and silica fume.

例如，具有高 CaO 含量的组 20 和组 23 代表**钙长石（$CaO \cdot Al_2O_3 \cdot 2SiO_2$）基隔热耐火砖**，最高温度限制分别为 1093 ℃和 1260 ℃。

For example, the groups 20 and 23 with high CaO content represent **the anorthite ($CaO \cdot Al_2O_3 \cdot 2SiO_2$) based insulating firebricks** with maximum temperature limits of 1093 ℃ and 1260 ℃, respectively.

在这项研究中，通过将回收的纸张加工废料和锯末添加到两种不同类型的黏土中，开发了**多孔钙长石隔热耐火砖**。

In this study, **porous anorthite insulating firebricks** have been developed by adding recycled paper processing wastes and sawdust to two different types of clay.

研究了三种不同黏土如富集黏土、商业黏土和耐火黏土对制造**钙长石基轻质耐火砖**的适用性。

Suitability of three different clays such as enriched clay, commercial clay and fireclay for manufacturing of **anorthite based lightweight refractory bricks** was studied.

石棉不能归类为耐火材料，尽管通常它的性能足以承受过热蒸汽和其他高温工业环境。

Asbestos cannot be classed as refractory, although normally its properties are sufficient to withstand superheated steam and other elevated temperature industrial environments.

岩棉是通过将白云石、玄武岩和辉绿岩等石头在 1400～1600 ℃下转变为纤维而制成的。

Rock wool is produced by turning stones such as dolomite, basalt, and diabase into the fiber at 1400-1600 ℃.

普通硅酸铝纤维喷涂材料和高纯**硅酸铝纤维**喷涂材料的最高工作温度分别为700 ℃和1200 ℃，因此安全工作温度分别为630 ℃和1080 ℃。

The highest operating temperature of both common **aluminium silicate fibre** spray materials and high-purity aluminium silicate fibre spray materials are 700 ℃ and 1200 ℃ respectively, so the safe operating temperature are 630 ℃ and 1080 ℃ respectively.

硅酸铝纤维增强壳体的烧成强度高于其他材料。

The fired strength of shell reinforced with **aluminium silicate fibre** is higher than that of the others.

硅酸铝耐火纤维材料具有良好的耐热性、绝热性、质量轻等优点，广泛应用于化学、冶金、机械制造等领域。

Alumina-silicate refractory fiber materials are widely used in chemistry, metallurgy, mechanical manufacture fields, which have the advantages of good heat resistance, good heat insulation, light mass etc.

先前对**多晶氧化铝纤维** Nextel™ 610 的研究表明，在高达 1200 ℃的温度和10 h 的时间内，几乎没有晶粒生长。

Prior work on the **polycrystalline alumina fiber** Nextel™ 610 demonstrated that little grain growth occurred at temperature up to 1200 ℃ and time of 10 h.

商用**多晶氧化铝纤维**（Nextel™ 610，3M）的加工温度限制在 1200 ℃，这样是为了避免显著的晶粒生长，从而降低强度。

Commercially available **polycrystalline alumina fibers** (Nextel™ 610, 3M) are limited to processing temperature of 1200 ℃ to avoid significant grain growth that can lead to lower strength.

将**氧化锆纤维**浆料与二氧化硅黏结剂混合形成均匀浆料，形成**纤维状多孔氧化锆陶瓷**。

Fibrous porous zirconia ceramics were formed by mixing the slurry of **zirconia fibers** and a silica binder to form uniform slurry.

多孔纤维网络中**氧化锆纤维**含量低是导致纤维状多孔陶瓷导热系数略有下降的原因之一。

The low **zirconia fiber** content of porous fiber networks is actually one reason why the thermal conductivity of fibrous porous ceramics was slightly decreased.

超细二氧化硅粉末多年来一直被用作油墨、油漆和聚合物的触变添加剂、润滑剂的增稠剂、粉末的流动控制剂、电子、催化剂基材的增强填料以及高级玻璃的原料。

Ultrafine silica powders has been applied as thixotropic additives inks, paints and polymers, thickening for greasers, flow control agents for powders, reinforcing fillers for electronic, catalyst substrates and as raw materials for advanced glass for many years.

为了研究**多孔硅酸钙**作为温度高达 1100 K 的参考材料的可行性，GEFTA 在内的德国热物理工作组与 7 个参与者一起对一种商用**硅酸钙隔热材料**的导热系数进行了对比。

In order to investigate **porous calcium silicate** as reference material for temperatures up to 1100 K, the German Thermophysics Working Team within GEFTA initiated an intercomparison of thermal conductivity measurements on a commercially available **calcium silicate insulation material** with seven participating laboratories.

为了提高钢锭铸造热顶用**隔热板**的隔热性能，采用数值模拟和量热法研究了不同孔隙率的固体结构和多孔结构保温板的导热系数。

In order to improve heat prevention property of **insulation board** used in hot top during casting of steel ingot, thermal conductivities of insulation boards of solid structure and porous structure with different porosities were investigated using numerical simulation and calorimetric techniques.

在这里，我们报告了一种具有锯齿形结构的**亚晶锆石纳米纤维气凝胶**的多尺度设计，可在高温下实现出色的热机械稳定性和良好的隔热性。

Here we report a multiscale design of **hypocrystalline zircon nanofibrous aerogels** with a zig-zag architecture that leads to exceptional thermomechanical stability and ultralow thermal conductivity at high temperatures.

陶瓷气凝胶是一种很有潜力的隔热材料，但其机械稳定性差，在热冲击下容易降解。

Ceramic aerogels are attractive for thermal insulation but plagued by poor mechanical stability and degradation under thermal shock.

研究了 1600 ℃烧结**微孔镁基耐火材料**的显微组织、孔径分布、相组成、热膨胀系数、导热系数和烧结性能。

The microstructure, pore size distribution, phase composition, coefficient of thermal expansion, thermal conductivity and sintering properties of the **microporous magnesia based refractory products** sintered at 1600 ℃ were characterized.

因此，大量的研究集中在更精确地控制**轻质耐火材料**中孔隙的大小和分布，并获得含有微孔的轻质耐火材料。

Therefore, numerous researches focused on more accurate control to the size and the distribution of the pores in the **lightweight refractory**, and the lightweight refractory containing micropores was obtained.

节能涂料改善了格子砖在热风炉中的表面辐射性能，加强了燃烧期间气体与格子砖之间的辐射传热过程，增加了格子砖的蓄热能力和表面温度，可以间接强化鼓风期对流换热。

Energy-saving coating improves the surface radiation property of checker bricks in hot stove which strengthens the radiation heat transfer process between gas and the checker bricks during the combustion period which increases the heat storage capacity and the surface temperature of the checker bricks which can indirectly strengthen the convection heat transfer during blast period.

得出结论，**$LaMnO_3$-磷酸盐涂层**作为高温工业炉**节能涂层**具有良好的应用前景。

It is concluded that the **$LaMnO_3$-phosphate coating** has good application prospects as an **energy-saving coating** for high-temperature industrial furnaces.

与没有涂层相比，在电阻炉中使用这种**高温红外辐射节能涂层**可显著节省能源。

Using this **high temperature infrared radiation energy saving coating** in a resistance furnace resulted in significant energy savings compared to no coating.

5.7.3 术语段落

现已确定，可以使用大量的废料来制造**隔热耐火材料**。研究了掺入飞灰、稻壳灰、稻壳和烧成耐火熟料的绝热试件的性能。利用各种废弃物作为生产隔热耐火材料的原料，可成为这些大量废弃物回收利用的一种重要途径。通过改变混合料中 FA 的含量，可以改善烧结体的隔热性能。结果表明，该材料具有低导热系数、高气孔率、低密度、较好的电阻率、中等的 CCS 和抗折强度。在 800 ℃下烧制的样品 s-6 可以被认为是用于隔热目的的耐火材料。研究结果表明，该方法在大规模合成保温砖方面具有广阔的应用前景[42]。

It is established that substantial amounts of waste materials can be used to fabricate **insulation refractory**. The properties of insulation samples incorporating fly ash, rice husk ash, rice husk and fired refractory grog are investigated. Use of different wastes as a raw material in fabrication of insulation refractories can be a significant way of recycling for final disposal of these abundant wastes. Varying the FA content in mixtures it is possible to improve the insulation properties of sintered bodies. Results show excellent low thermal conductivity, high porosity, low density, good electrical resistivity, moderate CCS and flexural strength. Sample s-6 fired at 800 ℃ can be envisaged as a refractory material to be used for insulation purpose. The obtained results show that it would be promising for the large-scale synthesis of insulation bricks[42].

耐火材料按气孔率可分为致密耐火材料和多孔耐火材料两大类。致密耐火材料的残余气孔率小于 20%，用于与热熔体和气体直接作用。多孔耐火材料具有比其他耐火材料更低的热容量和导热系数，一般用于不同熔炉或窑炉的隔热。这些**轻质隔热耐火材料**有助于提高上部结构的热效率，大幅降低燃料消耗。这些耐火材料主要是用天然原料如膨胀蛭石、高岭土、硅酸钙、珍珠岩、石英、硅藻土和其他轻质耐火骨料按传统方法合成的。耐火材料工业对天然原材料的大量消耗造成了这些自然资源的短缺[43]。

The refractories could be categorized into two groups according to the porosity, that is, dense and porous refractories. The retained porosity of dense refractories is less than 20%, and it is used in direct interaction with the hot molten matter and gases. The porous refractories have a much lower heat capacity and thermal conductivity than other refractories. Generally, it is used for heat insulation in different furnaces or kilns. These **lightweight insulating refractories** help to increase thermal efficiencies in the superstructure, reducing fuel consumption substantially. These refractories are primarily synthesized using natural ingredients like expanded vermiculite, kaolin, calcium silicate, perlite, quartz, diatomite, and other light-weight refractory aggregates through conventional route. The massive consumption of natural raw materials in the refractory industries has created a shortage of these natural resources[43].

在水培温室中，研究了**膨胀蛭石**与营养液长期接触后水物理性质的变化。我们的研究结果表明，在水培温室中长期使用热蛭石的过程中，会发生机械破坏和矿物成分变化。在水培温室中，含有高浓度磷灰石的蛭石会发生变化，这些变化是在向植物提供足够的钾的条件下自然风化过程中观察到的。在这方面，它不会对自然环境构成危险。废蛭石的化学成分中重金属的累积量不超过标准值[44]。

Changes in the hydro-physical properties of **expanded vermiculite** as a result of prolonged contact with nutrient solutions in a hydroponic greenhouse were studied. The results of our study showed that both mechanical destruction and changes in mineral composition occur during the long-term operation of thermo-vermiculite in a hydroponic greenhouse. In a hydroponic greenhouse, vermiculite with a high concentration of phlogopite undergoes changes that are observed in the process of natural weathering under conditions of sufficient provision of plants with potassium. In this regard, it does not pose a danger to the natural environment. The accumulation of heavy metals in the chemical composition of spent vermiculite does not exceed standard values[44].

近年来，密度非常低的非结构轻质骨料，如蛭石、**珍珠岩**和膨胀聚苯乙烯骨料，被用作水泥基砂浆中的部分沙子替代物。这些材料的主要优点是由于其高度多孔的性质而具有良好的隔热性能。文献中证明了掺入这些骨料的水泥基材料的隔热性能得到改善。基于此，由于这些材料的多孔结构，也有可能在抹灰砂浆中使用这些材料以增强对高温的抵抗力。虽然**膨胀珍珠岩**的使用更为广泛，但关于蛭石在水泥基材料中的利用的研究相当有限，直到最近才受到越来越多的关注。蛭石是水合镁铝铁硅酸盐，由云母蚀变而成，呈薄片状。当通过加热到 900 ℃或更高的温度进行剥离时，水会释放出来，薄片会膨胀成非常轻的多孔材料。由此所得的**膨胀蛭石**被认为具有耐热性并表现出良好的隔音和隔热性能[45]。

In recent times, non-structural lightweight aggregates such as vermiculite, **perlite** and expanded polystyrene aggregates which are very low in density were incorporated as partial sand replacement in cement-based mortars. The prime advantage of these materials is the good insulation properties owing to their highly porous nature. The improved thermal insulation performance of cement-based materials incorporating these aggregates were evidenced in literatures. Based on this, because of the porous structure of these materials, there is also potential in utilizing these materials in plastering mortars for enhanced resistance towards elevated temperature. While the use of **expanded perlite** is more widespread, research on the utilization of vermiculite in cement-based materials is fairly limited and only recently garnering increased attention. Vermiculite is hydrated magnesium-aluminium-iron silicate, formed by the alteration of mica and appear in the forms of flakes. When exfoliated by heating to temperature of 900 ℃ or higher, water is released and the flakes expand into very lightweight porous material. The resulting **expanded vermiculite** is considered to be heat resistant and exhibits good sound and thermal insulation properties[45].

硅藻土是单细胞硅藻沉积的外骨骼，是一种很有前途的天然材料。硅藻土是由具有独特三维结构的无定形二氧化硅和其表面的硅醇基团组成的，使其易于功能化。由于硅藻土独特的物理特性，如高比表面积、低密度、高孔隙率、优异的机械强度和无毒，硅藻土可用于增强填料、催化剂和药物输送。一些研究人员发现，在硅藻土中负载吲哚美辛可以延长药物释放 2 周以上，硅藻土负载的铂纳米颗粒具有很高的催化剂活性和可重复使用性。此外，在生物材料中，硅藻土已被用于增强骨组织工程应用的壳聚糖膜的力学性能和生物活性。作为一种增强填料，硅藻土价格相对较低，而且带负电荷。因此，硅藻土可以用于药物输送的水凝胶中[46]。

Diatomite as a promising natural material is the deposited exoskeletons of the single-celled diatoms. Diatomite is composited by amorphous SiO_2 with a unique three-dimension architecture and the silanol groups on its surface, making it easy for functionalization. Given the diatomite′s unique physical characteristics, such as high specific surface area, low density, high porosity, excellent mechanical strength, and nontoxicity, diatomite can be used for reinforcement fillers, catalysis, and drug delivery. Some researchers found that indomethacin loaded in diatomite can prolong drug release over 2 weeks, and diatomite-supported Pt nanoparticles have high catalyst activity and reusability. Moreover, in biomaterials, diatomite has been used to reinforce the mechanical property and improve the biological activity of chitosan membranes for bone-tissue engineering applications. As a reinforcing filler, diatomite has a relatively low price and is negatively charged. Thus, diatomite can be used in hydrogels for drug delivery[46].

粉煤灰是煤燃烧后烟气产生的细灰，是典型的硅酸盐固体废物。在新型陶瓷开发中使用**漂珠**提高了粉煤灰的利用价值。考虑到漂珠优异的物理化学性能，烧成的陶瓷表现出良好的特性。粉煤灰还可用于制备陶瓷墙砖的釉料[47]。

Fly ash is fine ash generated from flue gas after coal combustion and is a typical silicate solid waste. Using **floating bead** in the development of new ceramics improves the utilization value of fly ash. When the excellent physical and chemical properties of floating beads are considered, ceramics produced by firing show good characteristics. Fly ash can also be used to prepare glaze for ceramic wall tiles[47].

粉煤灰漂珠是铝硅酸盐无机粉末材料，从**粉煤灰**中提取，主要由 SiO_2 和 Al_2O_3 组成，一般呈灰色，显微镜下呈透明的空心玻璃球。粉煤灰漂珠具有质轻、耐磨、反光、抗辐射、耐高温、耐酸碱、自润滑等多种特性。粉煤灰漂珠广泛应用于橡胶、塑料、黏合剂、涂料和功能材料等领域。由于粉煤灰漂珠在聚合物中的分散性和粉煤灰漂珠与聚合物的界面附着力较差的物理和化学特性，这可能导致两相分离。因此，**粉煤灰漂珠**需要进行改性，以提高与聚合物的相容性[48]。

Fly ash floating beads were aluminosilicate inorganic powder materials, extracted from **fly ash**, mainly composed of SiO_2 and Al_2O_3, generally grey in colour, transparent hollow glass spheres under the microscope. Fly ash floating beads had a variety of features, such as light mass, wear resistance, reflective, radiation resistance, high temperature resistance, acid and alkali resistance, self-lubricating, etc. Fly ash floating beads were widely used in rubber, plastics, adhesives, coatings and functional materials and other fields. Due to the physical and chemical characteristics of fly ash floating beads, the dispersion of fly ash floating beads in polymer and the interfacial adhesion between fly ash floating beads and polymer were poor, which might result in two-phase separation. Therefore, **fly ash floating beads** needed to be modified to improve the compatibility with polymer[48].

形态稳定相变材料（FSPCMs）是用石蜡和**氧化铝空心球**制备的。由熔融阶段的峰值相变温度和潜热决定的 FSPCM 的热性能分别为 23.6 ℃和 69.8 J/g，凝固阶段分别为 15.8 ℃和 71.3 J/g[49]。

Form-stable phase change materials (FSPCMs) were prepared with paraffin and **alumina hollow spheres**. The thermal properties of FSPCM determined by the peak phase transition temperature and latent heat in the melting stage were 23.6 ℃ and 69.8 J/g, respectively, and those in the solidification stage were 15.8 ℃ and 71.3 J/g, respectively[49].

为了高效催化难降解废水中木质素的降解，采用“溶胶-凝胶+煅烧+真空浸渍”三种模板法制备了一种新型纳米复合材料，Ru 纳米颗粒嵌入 **$TiZrO_4$ 空心球**表面，$TiZrO_4$/Ru 独特的二元组成阻止了 Ru 的聚集并保持其高活性。在 160 ℃和 2. 0 MPa 氧气下催化氧化 3 h，在 $TiZrO_4$/Ru 的催化下，98%的碱木质素被降解，70%的有机碳被矿化，而在没有分析剂的情况下，这些值只有 50%和 25%。通过 EPR 测定，由于高效羟基自由基的产生，该催化剂将碱性木质素的催化氧化速率常数 k_1（h^{-1}）从 0. 282 h^{-1}提高到 1. 175 h^{-1}[50]。

To catalyze the degradation of lignin in refractory wastewater efficiently, a new nanocomposite with Ru nanoparticles embedded on the surface of **$TiZrO_4$ hollow spheres** was fabricated with three method a "sol-gel + calcination + vacuum-impregnation" template method, and the unique binary composition of $TiZrO_4$/Ru prevented the aggregation of Ru and keep its high activity. During 3-h catalytic-oxidation at 160 ℃ and 2. 0 MPa O_2, 98% alkali lignin was degraded and 70% organic carbon was mineralized with the catalysis of $TiZrO_4$/Ru, while the values were only 50% and 25% without analysts. The catalyst increased the catalytic-oxidation rate constant k_1 (h^{-1}) of alkali lignin from 0. 282 h^{-1} to 1. 175 h^{-1} because of high-efficiency hydroxyl radical production, as determined by EPR[50].

添加**硅藻土**提高了水泥浆的强度，与一级硅藻土混合的水泥浆的强度高于与一级或二级硅藻土混合的水泥浆的强度；煅烧一级硅藻土和二级硅藻土的最佳用量均为 5%；与空白试件相比，水泥浆试件的抗压强度分别提高了 54. 6%、15. 4%和 15. 4%[51]。

Adding **diatomite** improved the strength of cement paste, the strength of the cement paste mixed with the first-grade diatomite is higher than that of the cement paste mixed with first-grade or second-grade diatomite; the optimal dosage of calcined first-grade diatomite, first-grade diatomite and second-grade diatomite all are 5%; compared with the blank specimen, the compressive strength of cement paste specimen has increased by 54. 6%, 15. 4% and 15. 4% respectively[51].

随着许多国家开始转向可再生能源，世界各地燃烧烟煤时产生的传统硅质 F 级**粉煤灰**的供应量正在减少。现有的燃煤电厂也开始将烟煤转化为次烟煤或将两者混合以减少排放和成本，前者会产生钙质 C 级粉煤灰。为了延长**硅质粉煤灰**的供应，供应商已开始分销混合煤灰，并开始将硅质粉煤灰与其他燃煤产品（CCPs）混合，例如**钙质粉煤灰**或底灰，又或是天然火山灰。在这项研究中检查了这些混合粉煤灰，并与传统的硅质粉煤灰进行了比较。尽管这些粉煤灰具有与传统硅质粉煤灰相似的氧化物成分和物理特性，但粉煤灰中存在的结晶相和玻璃相可能不同。这会改变水泥水化并导致由于硫酸盐侵蚀而导致的膨胀无法缓解。在本研究中进行的所有其他性能和耐久性测试中，所有粉煤灰在水泥基混合物中的表现都很好[52]。

The availability of traditional siliceous Class F **fly ash**, produced when burning bituminous coals, is decreasing around the world as many countries begin to switch to renewable energy sources. Existing coal burning power plants are also starting to convert from bituminous to subbituminous coals, which produce calcareous Class C fly ash, or to a blend of the two to reduce emissions and cost. To prolong the availability of **siliceous fly ash**, suppliers have started distributing blended coal ashes and have also begun to blend siliceous fly ash with other coal combustion products (CCPs), such as **calcareous fly ash** or bottom ash, or natural pozzolans. A variety of these blended fly ashes were examined in this study and compared to a traditional siliceous fly ash. Although these fly ashes have similar oxide compositions and physical characteristics to a traditional siliceous fly ash, the crystalline and glassy phases present in the fly ash may differ. This can alter cement hydration and result in failure to mitigate expansion due to sulfate attack. In all other performance and durability tests conducted in this study, all fly ashes performed well in cement-based mixtures[52].

采用直接混合法制备石蜡/膨胀珍珠岩水泥砂浆。研究了其力学性能和热性能。本研究得出以下结论：石蜡与**膨胀珍珠岩**具有良好的相容性。石蜡/膨胀珍珠岩材料中石蜡的平均含量（质量分数）高达65%。石蜡/膨胀珍珠岩材料只是一种物理相互作用。水泥砂浆的抗压强度和抗弯强度随着石蜡/膨胀珍珠岩材料用量的增加而降低。水泥砂浆用石蜡/膨胀珍珠岩材料的最佳质量含量确定为20%（水泥）。石蜡/膨胀珍珠岩材料水泥砂浆的储热/释放曲线在100次循环后略有变化。含有20%石蜡/膨胀珍珠岩材料的水泥砂浆具有良好的蓄热性和热稳定性[53]。

Cement mortar with paraffin/**expanded perlite** materials was prepared by direct mixing method. Mechanical and thermal properties were investigated. The following conclusions are drawn in this study: Paraffin has a good compatibility with expanded perlite. The average content (mass fraction) of paraffin in paraffin/expanded perlite material reaches as high as 65%. The paraffin/expanded perlite material is just a physical interaction. The compressive and flexural strength of cement mortar decrease with increasing amount of paraffin/expanded perlite materials. The optimal mass content of paraffin/expanded perlite material for cement mortar is determined as 20% (of cement). The heat storage/release curves of cement mortar with paraffin/expanded perlite materials changed slightly after 100 cycles. Cement mortar with 20% paraffin/expanded perlite materials has good heat storage and thermal stability[53].

氧化铝轻质隔热材料有孔隙率高的特点，因此具有较低的导热系数，氧化铝还具有较小的固有温度，耐腐蚀和良好的化学稳定性，适用于诸多领域。目前轻质砖和浇注料的骨料氧化铝空心球是**氧化铝轻质隔热制品**的主要形式。这些产品具有强度高、质量轻、隔热性好等优点，可替代传统的热窑部件材料[54]。

Alumina lightweight insulation materials is characterized by high porosity, thus has a lower coefficient of thermal conductivity, alumina also possesses small inherent temperature, corrosion resistance and good chemical stability, which make it applicable in many fields. Currently aggregate alumina hollow ball of light brick and castable are the main forms of **alumina lightweight insulation products**. The good properties of these products, such as high strength, light mass and good thermal insulation, enable their replacement of traditional hot kiln section materials[54].

在这项研究中，研究了由不同类型的黏土（K244 黏土和耐火黏土）、再生纸张加工废料和添加锯末的混合物生产**多孔轻质钙长石基隔热陶瓷**。得出的结论是，回收的纸张加工废料由于其有机和无机含量，可用作生产多孔钙长石陶瓷的合适替代原料来源。来自废纸的氧化钙和来自黏土的硅酸铝之间的反应导致钙长石（$CaO \cdot Al_2O_3 \cdot 2SiO_2$）与少量第二相的形成。还评估了黏土 244 与碱和耐火黏土在制造**钙长石基轻质隔热陶瓷**中的适用性。在具有钙长石成分的混合物中添加锯末有助于增加样品的孔隙率[55]。

In this study, the production of **porous and lightweight anorthite based insulating ceramics** from mixtures of different types of clay (K244 clay and fireclay), recycled paper processing waste and sawdust addition was investigated. It was concluded that the recycled paper processing wastes could be used as a suitable alternative raw material source for production of porous anorthite ceramics due to their organic and inorganic content. The reaction between calcium oxide from paper waste and aluminum silicate from clays resulted in the formation of anorthite ($CaO \cdot Al_2O_3 \cdot 2SiO_2$) with a minor secondary phase. Suitability of clay 244 with alkalis and fireclay in the manufacturing of **anorthite based lightweight insulating ceramics** was also evaluated. Sawdust addition into the mixtures with anorthite composition contributed to increase porosity of the samples[55].

这项研究是在另一项涉及生产用于建筑物的垂直多孔隔热陶砖的研究之后进行的。由碳酸钙和纤维素组成的造纸加工残渣被用作黏土原料的添加剂来制造砖。在该研究中，在黏土材料中添加了不同数量的纸渣，发现过量添加会在砖中产生钙长石和钙铝黄长石。因此，激发了制造具有多孔钙长石砖的想法。碳酸钙含量和细纤维素纤维都有助于制造**多孔钙长石基陶瓷**。本研究中生产的多孔陶瓷可在高达 1200 ℃的温度下用作工业和实验室用工业熔炉和电窑的备用隔热耐火材料。这些耐火材料中的孔隙率可以高达 60%[56]。

This study followed another study which dealt with the production of vertically perforated insulating earthenware brick for use in buildings. Paper processing residues composed of calcium carbonate and cellulose were used as an additive to clay raw material to make brick. In that study, different amount of paper residue additions were made to clay material and it was found that excessive additions produced anorthite and gehlenite in the brick. So the idea of making anorthite brick with large amounts of porosity was inspired. Not only its calcium carbonate content but also its fine cellulose fibers helped in making **porous anorthite based ceramics**. Porous ceramics produced in this study could find applications up to 1200 ℃ as refractories in backup insulation for industrial furnaces and electrical kilns for industrial and lab use. Amount of porosity in these refractories can be up to 60%[56].

这项研究的重点是**多孔轻质黏土砖**作为建筑材料，而不是蒸压加气混凝土和浮石块。传统黏土砖在自然条件下表现出良好的稳定性，与其他建筑材料相比是一种经济的替代品。此外，黏土砖的力学性能比蒸压加气混凝土和浮石块好，但保温性能较差，因此它们在建筑物中的使用变得越来越少。本研究的重点是使用复制方法开发具有相对更好的隔热性能的多孔轻质黏土砖[57]。

This study focused on **porous lightweight clay bricks** as a construction material as opposed to autoclaved aerated concrete and pumice blocks. Traditional clay bricks exhibit good stability against environmental conditions and are an economical alternative compared to other construction materials. Furthermore, clay bricks have better mechanical properties than autoclaved aerated concrete and pumice blocks but poorer thermal insulation properties; hence, their use in buildings is becoming increasingly unfavourable. This study focused on developing porous lightweight clay bricks with relatively better thermal insulation properties using a replication method[57].

温石棉含有 14%（按质量计）的羟基（—OH）基团，这些羟基在温度高于 450 ℃时从其结构中丢失。这种水分蒸发的潜热被认为是一种有效的吸热剂，保护了剩余的未降解纤维。此外，固体分解产物是惰性的，导热性低，为剩余的纤维提供额外的保护，并保持结构完整性。已经证明，在某些情况下，**石棉**可以在高达 1700 ℃的温度下保持其完整性[58]。

Chrysotile contains 14% by mass of hydroxyl (—OH) groups, which is lost from its structure at temperatures greater than 450 ℃. The latent heat of vaporisation of this water content is thought to be a potent heat sink, protecting the remaining undegraded fibre. Further, the solid decomposition products are inert and of low thermal conductivity, providing additional protection to the remaining fibres, and maintaining structural integrity. It has been shown that, in some cases, **asbestos** can maintain its integrity at temperatures up to 1700 ℃[58].

图 11 和图 12 分别为锥形量热测试后**石棉酚醛复合材料**和 NKA_3 纳米复合材料烧蚀试样的扫描电镜图。这些图显示了顶面（焦区表面）、侧面和热流方向。烧蚀区、原始材料、多孔反应区和致密炭层的特征在所有样品中都很明显[59]。

Figs. 11 and 12 show scanning electron micrographs of the **asbestos-phenolic composite** and NKA_3 nanocomposite ablative sample after cone calorimetry test, respectively. These figures are illustrating top surface (surface of char region), lateral surfaces, and heat flux direction. The characteristics of ablation regions, virgin material, porous reaction zone and dense char layer are apparent in all of the samples[59].

新型热回收锅炉保温结构采用纤维喷雾保温方案，其结构由高纯**硅酸铝纤维**和普通硅酸铝纤维组成。由于保温施工内部与炉内的烟气和火焰直接接触，其温度较高，因此内部采用高纯硅酸铝纤维喷涂的耐高温保温材料，外部采用普通硅酸铝纤维涂层，以降低保温材料成本[60]。

New thermal insulation construction of thermal recovery boiler use fiber spray thermal insulation program, its structure is consisted of high-purity **aluminum silicate fiber** and common aluminum silicate fiber. Due to the inner of thermal insulation construction contact with the smoke and flame inside the furnace directly whose temperature is higher, so the inner use high-purity aluminum silicate fibre spray thermal insulation material of high-temperature resistance, and the outer use common aluminum silicate fiber coating to reduce thermal insulation material cost[60].

另一种具有莫来石氧化铝成分的 85% Al_2O_3-15%二氧化硅纤维的抗蠕变性优于**多晶氧化铝纤维**。这引发了商用氧化铝-二氧化硅纤维的开发，例如 3M 的

Nextel™ 720 纤维。据推测，在高温环境下，不稳定的无定形 Altex 纤维微观结构也会转变为更稳定、抗蠕变的含有莫来石的微观结构[61]。

Another 85% Al_2O_3-15% silica fiber with a mullitealumina composition was shown to have creep resistance superior to that of the **polycrystalline alumina fibers**. This led to the development of commercially available alumina-silica fibers such as 3M′s Nextel™ 720 fiber. Presumably, during high-temperature exposure, the unstable, amorphous Altex fiber microstructure also converts to a more stable, creep-resistant mullite-containing microstructure[61].

多晶氧化铝纤维是一种具有非凡特性的无机纤维材料，其内部结构为微晶态。具有热稳定性（可在 1700 ℃下使用）、耐腐蚀、导热率低、强度高等特点，广泛应用于耐高温材料和高功能复合材料增强纤维等领域[62]。

Polycrystalline alumina fiber is a kind of inorganic fiber material with extraordinary features, whose internal structure is microcrystal state. For the features like thermo stability (can be used at 1700 ℃), corrosion resistance, low heat conduction rate, high strength, etc. it is widely used in many fields as the high temperature resistance material and the high functional composite material strengthened fiber[62].

通过混合**氧化锆纤维**浆料和黏合剂、真空成型和烧结干燥毡的过程，制造了具有超高孔隙率（72%~89%）的**纤维氧化锆陶瓷**。考察了黏结剂含量和烧结温度对组织、线收缩率、热性能和力学性能的影响。结果表明：氧化锆纤维主要在纤维连接处黏结成功，10%的黏结剂和 1600 ℃是获得较好性能的最佳工艺参数。材料的室温导热系数和机械强度分别随孔隙率呈线性和指数变化。烧结陶瓷在 1600 ℃加热 1 h 后线收缩率小于 2%，密度低至 0.67~1.72 g/cm^3，导热系数低至 0.056~0.16 W/(m·K)，抗压强度较高，为 0.6~13.3 MPa。此外，还研究了该材料的高温（高达 1400 ℃）隔热性能[63]。

Fibrous zirconia ceramics with ultra-high porosity (72%-89%) were fabricated through a process of mixing the slurry of **zirconia fibers** and a binder, vacuum-molding and sintering the dried felt. The effects of binder content and sintering temperature on the microstructure, linear shrinkage, thermal and mechanical properties were investigated. The results showed that the zirconia fibers were bonded successfully mainly at the fiber junctions, and 10% binder and 1600 ℃ were the optimal processing parameters for better properties. It was also found that the materials′ room temperature thermal conductivity and mechanical strengths varied with porosity linearly or exponentially, respectively. The sintered ceramics had a linear shrinkage of less than 2% after heating at 1600 ℃ for 1 h, a low density of 0.67-1.72 g/cm^3, low thermal conductivity of 0.056-0.16 W/(m · K), and relatively high compressive strength of 0.6-13.3 MPa. In addition, the material's high-temperature (up to 1400 ℃) thermal insulation performance was also studied[63].

表1显示了无机黏结剂含量对**纤维状多孔氧化锆陶瓷**孔隙率和密度的影响。当B_4C含量（质量分数）从0%增加到4%，SiC含量恒定（10%）时，纤维状多孔氧化锆陶瓷的密度从0.44 g/cm^{3}增加到0.52 g/cm^3，孔隙率从92.8 g/cm^{3}下降到91.5%。结果表明，无机黏结剂对多孔陶瓷的密度有较大的影响。不同无机黏结剂含量的多孔氧化锆纤维网络的SEM图像如图1所示。由图1a可知，氧化锆纤维骨架结构具有各向异性。**氧化锆纤维**在*x/y*方向上分布相对均匀，在*z*方向上形成准层状结构。这可能是由于在成型过程中，在毛毡上施加了初始压力，以防止纤维骨架的重新结合和分层[64]。

Table 1 shows the effect of inorganic binder content on porosities and densities of **fibrous porous zirconia ceramics**. When the B_4C content (mass fraction) was increased from 0 to 4% with a constant content (mass fraction) of SiC (10%), the density of the fibrous porous zirconia ceramics increased from 0.44 g/cm^3 to 0.52 g/cm^3 and the porosity decreased from 92.8% to 91.5%. It indicates that the inorganic binder has an increased effect on the density of porous ceramics. The SEM images of the porous zirconia fiber networks with different inorganic binder contents were showed in Fig. 1. It can be seen in Fig. 1a that the zirconia fiber skeleton structure is anisotropic. **Zirconia fibers** were dispersed relatively uniformly in the *x/y* direction, and a quasi-layered structure was formed in the *z* direction. It can be attributed to the fact that an initial pressure was applied on the felt to preventing the rebinding and layering of the fiber skeleton during moulding[64].

以石灰石和石英为前驱体，采用水热间歇法合成**硅酸钙保温材料**。本研究中研究的样品材料由德国CALSITHERM Silikatbaustoffe GmbH公司制备，产品名称为SILCAL 1100。该**高多孔硅酸钙**的主要成分为46%～47%的CaO和44%～45%的SiO_2，形成了三维结晶骨架。典型的有效孔径在微米范围内[65]。

Calcium silicate insulation materials are synthesized from the precursor limestone and quartz in a hydrothermal batch process. The sample material investigated in this work was prepared by CALSITHERM Silikatbaustoffe GmbH, Germany, and is commercially treated under the product name SILCAL 1100. The main components of this **highly porous calcium silicate** are 46%-47% CaO and 44%-45% SiO_2 which forms a three-dimensional crystalline backbone. Typical effective pore diameters are in the range of microns[65].

微孔硅酸钙是最好的无机轻质高温保温材料之一。它是由硅藻土、硅灰、粉煤灰、污泥、沸石、无定形 SiO_2 等 SiO_2 之间的水热反应形成的。该反应在 180 ℃ 左右形成托贝莫来石晶体。该反应的机理包括在 SiO_2 颗粒表面形成硅酸钙凝胶，凝胶向浆体分散，随后在 SiO_2 颗粒上吸附更多的 $Ca(OH)_2$ 晶体，凝胶向浆体反复分散。这个过程一直持续到 SiO_2 和 $Ca(OH)_2$ 之间的反应完成[66]。

Microporous calcium silicate is one of the best inorganic lightweight thermal insulation materials for high temperatures. It is formed by a hydrothermal reaction between SiO_2, from diatomaceous earth, silica fume, fly ash, sludge, zeolite, amorphous SiO_2, etc. The reaction results in the formation of tobermorite crystals at about 180 ℃. The mechanism of this reaction involves the formation of calcium silicate gel on the surface of SiO_2 particles, the dispersion of the gel to the slurry, the subsequent adsorption of more $Ca(OH)_2$ crystals on the SiO_2 particles and the repeated dispersion of the gel to the slurry. This continues until the reaction between SiO_2 and $Ca(OH)_2$ is complete[66].

本文报道了一种多尺度设计和合成具有之字形结构的**亚晶锆石纳米纤维**气凝胶（ZAGs），以实现近零（泊松比）ν 和近零（热膨胀系数）α 的优异热力学性能。这种 ZAGs 具有很高机械灵活性，在剧烈热冲击和高温暴露下的稳定性，以及卓越的高温隔热性能（1000 ℃时 10^4 mW/(m·K)，25 ℃和空气中 26 mW/(m·K)），在极端条件下提供可靠的隔热材料[67]。

Here we report a multiscale design and synthesis of **hypocrystalline zircon nanofibrous aerogels** (ZAGs) with a zig-zag architecture, to realize near-zero ν and near-zero α for superior thermomechanical properties. The resulting ZAGs feature high mechanical flexibility, high thermal stability under sharp thermal shocks and high-temperature exposures, and exceptional high-temperature thermal insulating performance (10^4 mW/(m·K) at 1000 ℃ and 26 mW/(m·K) at 25 ℃ and in air), presenting a reliable material for thermal insulation under extreme conditions[67].

因此，**陶瓷气凝胶**结合了低导热性和稳健的热稳定性，为暴露于极端条件下的热超绝缘提供了相当大的优势，而传统的**陶瓷气凝胶**在这种条件下聚合物和碳质绝缘材料很容易坍塌或点燃，如 SiO_2、Al_2O_3、SiC 和 BN 表现出较差的机械稳定性[68]。

Ceramic aerogels thus present a combination of low thermal conductivity and robust thermal stability that offers considerable advantage for thermal superinsulation exposed to extreme conditions under which polymeric and carbonaceous insulating materials could easily collapse or ignite, and the traditional **ceramic aerogels**, such as SiO_2, Al_2O_3, SiC, and BN, show poor mechanical stabilities[68].

传统的微孔耐火材料由于成本低，通常采用直接发泡、有机/无机物分解、原位成孔、反应键合技术、超塑发泡等方法制备，存在微观结构不均匀、不稳定等问题。合成的轻质集料具有高表观孔隙率和高强度的特点，可用于制备微孔耐火材料。采用两段烧结法制备了**微孔镁基耐火材料**。首先，在第一步加热中引入致气孔剂，合成结构均匀、性能稳定的轻质骨料；在此基础上，将不同粒径的轻质骨料制备微孔镁基耐火材料，并在第二步对其进行加热，得到符合预期环保标准的产品。重点研究了造孔剂对合成骨料的热质量和热流的影响，以及微孔镁基耐火材料的容重、表观孔隙率、热膨胀率、微观结构、孔径分布、导热系数、物相组成的影响[69]。

Due to the low cost, traditional microporous refractories that present nonuniform microstructure and instability are usually fabricated by direct foaming, decomposition of organic/inorganic matter, in situ pore forming, the react bonding technique, superplastic foaming et al. Synthesized lightweight aggregates, with the characteristic of high apparent porosity and high strength, can be used to produce the microporous refractories. The two-stage sintering method is used to prepare **microporous magnesia based refractory** in this paper. Firstly, the lightweight aggregates with uniform structure and stable performance are synthesized by introducing porogenic agent in the first heating step. Based on that, microporous magnesia based refractory is prepared by the lightweight aggregates of different particle sizes, and what's more they are heated in the second step to obtain the products in line with the expected environmental protection standards. The effects of porogenic agent on the thermal mass and heat flow of the synthesized aggregate, and the bulk density, apparent porosity, thermal expansivity, microstructure, pore size distribution, thermal conductivity, phase composition of the microporous magnesia based refractory products are emphasically investigated[69].

随着冶金工业的发展，钢包、中间包等高温容器用轻质保温材料在炼钢工业中发挥着越来越重要的作用，特别是具有良好保温性能的基础轻质耐火材料。氧化镁是碱性耐火材料的主要原料，因其具有良好的耐腐蚀性和高的耐载性，在炼钢工业中得到了广泛的应用。因此，开发和应用**轻质微孔镁基耐火材料**是十分重要的。孔隙的引入可以减少衬里材料在温度变化过程中的热胀冷缩，增强衬里材料的抗热剥落能力。轻质耐火材料一般由微孔骨料组成，其表观孔隙率可达45%，具有导热系数低、抗热冲击和耐腐蚀性能好等特点。在高温工业窑炉的使用过程中，导热系数低的内衬耐火材料在窑炉中表现出较好的保温性能[70]。

With the development of metallurgical industry, the lightweight heat-insulation materials for high temperature vessel including ladle, tundish and others have played an increasingly important role in the iron and steel-making industry, especially the basic lightweight refractories with good thermal insulation properties. Magnesia is a main raw material of basic refractories, which are widely applied in steel making industry due to good corrosion resistance and high refractoriness under load. Thus, the development and application of **lightweight microporous magnesia-based refractories** is extremely important. The introduction of pores can reduce the thermal expansion and shrinkage during the temperature changes, enhancing the thermal spalling resistance of the lining materials. Generally, lightweight refractories with low thermal conductivity coefficient, good thermal-shock and corrosion resistance are composed of microporous aggregates with apparent porosity up to 45%. During the service process of high temperature industrial furnaces, the lining refractories with low thermal conductivity show a better thermal insulation performance in furnace[70].

高发射率节能涂料由于能提高工业窑炉的热效率，一直被认为是先进的节能材料。对于高炉热风炉，方格砖涂覆该涂料可获得较高的风温。然而，在高炉热风炉上，涂料的最佳应用部位还不是很清楚，为此开展了**节能涂料**的优化应用研究[71]。

The high emissivity energy-saving coating is always considered to be advanced energy-saving material because it can increase thermal efficiency for industrial furnaces. For BF hot blast stoves, higher blast temperature can be obtained with checker bricks covered with the coating. However, the best application part of coating on BF hot blast stove is not well known, so the study of the optimization application of **energy-saving coating** was carried out[71].

第一步，通过固相反应工艺成功制备了一系列 $LaCr_xMn_{1-x}O_3$（x = 0, 0.25, 0.5, 0.75, 1）粉末。结果表明，制备的粉末的发射率随着 Mn^{3+} 含量的增加而增加，引起更强的分子振动。在制备的粉末中，$LaMnO_3$ 在 600 ℃时表现出最高的红外发射率（$\varepsilon_{3\text{-}5\,\mu m}=0.922$）。在下一步中，使用 $LaMnO_3$（作为填料）和磷酸盐（作为黏合剂）在 316L 合金和耐火砖基材上成功制备了**高发射率涂层**。所制备的涂料应用于圆柱形耐火砖内壁时表现出良好的节能性能（高达 13.2%）。

高温循环热处理结果表明，制备的涂层具有优异的耐高温性能。得出结论，**$LaMnO_3$-磷酸盐涂层**作为高温工业炉节能涂层具有良好的应用前景[72]。

In the first step, a series of $LaCr_xMn_{1-x}O_3$ (x = 0, 0.25, 0.5, 0.75, 1) powders were successfully prepared by solid-state reaction process. Results indicated that the emissivity of the prepared powders increases with the increase in Mn^{3+} content, inducing stronger molecular vibrations. Among the prepared powders, $LaMnO_3$ exhibited the highest infrared emissivity at 600 ℃ ($\varepsilon_{3\text{-}5\ \mu m}$ = 0.922). In the next step, **high emissivity coatings** were successfully prepared using $LaMnO_3$ (as the filler) and phosphate (as the binder) on 316L alloy and firebrick substrates. The prepared coating showed good energy-saving performance (as high as 13.2%) when it was applied on the inner wall of a cylindrical firebrick. High-temperature cyclic heat treatment results showed that the prepared coating has excellent high-temperature resistance. It is concluded that **the $LaMnO_3$-phosphate coating** has good application prospects as an energy-saving coating for high-temperature industrial furnaces[72].

远红外涂料是一种新型节能技术，已广泛应用于国家制造业的各个领域。在电厂锅炉中的应用取得了丰硕的成果。根据国外关于锅炉应用远红外涂料的参考资料，在蒸发量相同的情况下，锅炉中心燃烧部位应用远红外涂料可提高 10 ℃以上，节约 10%左右的煤炭消耗。本文将详细分析远红外涂料的节能特性[73]。

The far-infrared coating is a new energy-saving technology which has been widely used in every field of national manufacturing. The application in the power plant boiler has reached great achievements. According to the foreign references about the application of far-infrared coatings in the boilers, the application of far-infrared coatings in the boilers could rise more than 10 ℃ in the central flaming part under the same condition of evaporation capacity, and it saved about 10% coal consuming. The paper will analyze the energy-saving features of the far-infrared coatings in details[73].

5.8 熔铸耐火材料制品

5.8.1 术语词组

熔铸铝硅锆(AZS)耐火材料 fused-cast AZS refractories

铸造 AZS 耐火材料 casting AZS refractories

熔融氧化锆-氧化铝-二氧化硅耐火材料 fused

zirconia-alumina-silica material (AZS) refractory
熔铸氧化锆耐火材料 fused cast zirconia refractory
高氧化锆熔铸耐火材料 high zirconia fused-cast refractories
熔融莫来石颗粒 fused mullite particlcs
熔融莫来石粉末 fused mullite powder
熔铸刚玉制品 fused cast corundum refractory products
熔铸辉石刚玉耐火材料 fused-cast baddeleyite-corundum refractories

5.8.2 术语句子

该产品由**熔融氧化铝-氧化锆-二氧化硅**晶粒组成，基体主要为氧化铬。合成产物具有很好的热循环性能。

This product consists of **fused alumina-zirconia-silica** grain with a matrix largely of chromic oxide. The resultant body has very good thermal cycling properties.

采取了一种基本方法，以开发三维数学模型来预测**熔铸 AZS 耐火砖**中的热流和应力产生。

A fundamental approach has been taken, centered on the development of a three-dimensional mathematical model to predict heat flow and stress generation in **fused-cast AZS refractory blocks**.

熔融 Al_2O_3-ZrO_2-SiO_2（AZS）耐火材料由于温度分布不均匀和化学成分偏析，在冷却过程中容易产生裂纹缺陷。

Fused Al_2O_3-ZrO_2-SiO_2 (AZS) refractory is easy to generate crack defects in the cooling process because of non-uniform temperature distribution and chemical composition segregation.

目前在熔凝炉上使用的耐火材料有 **Al_2O_3-ZrO_2-SiO_2（AZS）熔铸材料**和 Cr_2O_3 基材料。

The refractories currently used in frit furnaces are **Al_2O_3-ZrO_2-SiO_2 (AZS) fused cast materials** and Cr_2O_3-based materials.

熔铸 Al_2O_3-ZrO_2-SiO_2（FC-AZS）被认为是陶瓷熔体中的“玻璃接触耐火材料”，用于核废料固定化。

Fused/cast Al_2O_3-ZrO_2-SiO_2 (FC-AZS) is being considered as " glass contact refractory" within ceramic melters, to be used for nuclear waste immobilization.

熔铸铝硅锆材料（AZS）是一种专用于玻璃炉的耐火材料。

Fused cast zirconia-alumina-silica material (**AZS**) is a kind of refractory which is exclusively used in glass furnace.

杨氏模量随温度的变化已经研究了两种**高氧化锆熔铸耐火材料**在 1700 ℃的热激发下的变化。

Young's modulus evolutions versus temperature have been studied for two **high zirconia fused-cast refractories** undergoing thermal solicitations up to 1700 ℃.

熔融莫来石颗粒为原料，ρ-Al_2O_3 和多晶硅废料为添加剂，淀粉为成孔剂，采用凝胶浇注法制备了孔隙率高、抗压强度高的多孔莫来石陶瓷。

The porous mullite ceramics with high porosity and compressive strength were prepared by gel-casting method using **fused mullite particles** as the raw materials, ρ-Al_2O_3 and polysilicon waste as the additives, and starch as the pore forming agent.

因此，利用不同形式的工业废料生产**熔铸铬铝锆耐火材料**和高铬耐火材料，可以降低成本，扩大应用范围。

Therefore, involvement of different forms of industrial waste in the manufacture of **fused cast chrome-alumina-zirconia** and high-chrome refractories makes it possible to reduce their cost and to expand the volume of application.

熔铸辉石刚玉耐火材料（FCBCRs）具有独特的性能，如缺乏渗透性孔隙，对矿物熔体具有极高的耐腐蚀性，可用于高温系统衬里。

Fused-cast baddeleyite-corundum refractories (FCBCRs) have unique properties such as a lack of permeable porosity and exceptionally high corrosion resistance to mineral melts that are useful for lining high-temperature systems.

5.8.3 术语段落

熔铸耐火材料的导热性和耐腐蚀性主要取决于其相结构。然而，基于相结构的**熔铸 AZS 耐火材料**的热导率与腐蚀速率之间的关系尚不清楚。本文旨在通过数值模拟方法和实验测试，探讨熔铸 AZS 耐火材料的物相组成与物理性能之间的关系，包括热导率和对液体玻璃的耐腐蚀性。这为研制低导热、高耐腐蚀的熔融 AZS 耐火材料提供了理论和技术基础[74]。

The thermal conductivity and corrosion resistance of fused-cast refractories are mainly determined by their phase structure. However, the correlation between thermal conductivity and the corrosion rate of **fused-cast AZS refractories** based on their phase structure is not clear. This paper aims to explore the relationship between the phase composition and physical properties of the fused-cast AZS refractories through numerical simulation methods and experimental tests, including thermal conductivity and corrosion resistance to liquid glass. This provides basic theory and technology for the fused AZS refractories with low thermal conductivity and high corrosion resistance[74].

在工厂规模上，**FC-AZS** 是通过在电弧炉（石墨电极）中熔化适量的分析试剂级（大于 99%）二氧化硅、氧化锆、氧化铝和纯碱来生产的，温度为 2200~2400 ℃。所产生的熔体根据所需的块的形状被浇注到“砂制模具”中。随后，模具非常缓慢地冷却，以获得没有热应力的产品[75]。

At plant scale, **FC-AZS** is produced by melting appropriate quantities of analytical reagent grade (better than 99%) silica, zirconia, alumina and soda ash in electric arc furnace (with graphite electrodes) at 2200-2400 ℃. The melt produced is cast into sand made moulds depending on the shape of the blocks desired. Subsequently moulds are cooled very slowly to achieve products free from thermal stresses[75].

由于氧化锆相的连通性，在氧化锆同素异向转变以下的**高氧化锆熔铸耐火材料**的情况下，这项工作首次证明了氧半渗透过程在耐火材料泡化过程中的作用。玻璃相仅起机械作用，使耐火材料在加热和冷却时均能承受单斜-四方相变而不开裂。该研究还证明了难熔铁的氧化态和温度对起泡机制的延迟起泡的作用。从工业角度来看，这些结果也很重要，因为它们可以提供减少甚至消除起泡现象的方法[76]。

This work demonstrates for the first time the role of the oxygen semipermeation process through refractory in the blistering process, in the case of a **high zirconia fused-cast refractory** below the allotropic transformation of zirconia, owing to the connectivity of the zirconia phase. The vitreous phase only plays a mechanical role, allowing the refractory to withstand the monoclinic-tetragonal phase transformation without cracking both on heating and cooling. The study also demonstrates the role of refractory iron oxidation state, and of temperature, on the delayed initiation of the blistering mechanism. These results are also important from an industrial point of view because they make it possible to provide ways for reducing or even canceling the phenomenon of blistering[76].

以**熔融莫来石颗粒**为原料，ρ-Al_2O_3 和多晶硅废料为黏结剂，淀粉为成孔剂，采用凝胶浇注法制备多孔莫来石陶瓷。此外，还研究了成孔剂含量对多孔莫来石陶瓷表观孔隙率、抗压强度、孔径分布、热分析和压降的影响。结果表明，制备的多孔莫来石陶瓷具有较高的表观孔隙率、抗压强度和透气性。通过添加多晶硅废料和 ρ-Al_2O_3 在原位形成莫来石相，使熔融莫来石颗粒结合，提高了多孔莫来石陶瓷的性能。随着可溶性淀粉含量的增加，试样的表观孔隙率由 49.50%增加到 62.67%，抗压强度由 4.98 MPa 降低到 1.35 MPa。孔径分布均匀，透气性高，完全满足高温烟气过滤应用的要求[77]。

In this study, porous mullite ceramics were prepared by gel-casting using **fused mullite particles** as the raw materials, ρ-Al_2O_3 and polysilicon waste as the binders, and starch as the pore forming agent. In addition, the effects of pore forming agent content on the apparent porosity, compressive strength, pore size distribution, thermal analysis, and pressure drop of porous mullite ceramics were studied. The results show that the as-prepared porous mullite ceramics have high apparent porosity, compressive strength, and gas permeability. Mullite phase formed in situ by adding polysilicon waste and ρ-Al_2O_3, which bonded the fused mullite particles and improved the performance of porous mullite ceramics. With increasing soluble starch content, the apparent porosity of the samples increased from 49.50% to 62.67%, and the compressive strength decreased from 4.98 MPa to 1.35 MPa. Furthermore, uniform pore size distribution and high gas permeability fully meet the requirements for the application of high temperature flue gas filtration[77].

矿棉生产过程中熔体的高腐蚀性要求在熔炼炉衬中使用具有高耐腐蚀性的耐火材料。最有效的材料是熔融铸造含铬耐火材料，即铬-氧化铝-氧化锆 KhATs-30 和 KhTs-45，以及高铬耐火材料 KhPL-85 和 KhMG。然而，使用纯原料生产这些等级的含铬耐火材料是其高成本的原因之一。因此，利用不同形式的工业废料生产**熔铸铬铝锆耐火材料**和高铬耐火材料，可以降低成本，扩大应用范围[78]。

The high corrosiveness of melts during manufacture of mineral wool requires use of refractories with high corrosion resistance in melting furnace linings. The most effective materials, making it possible to increase a mineral wool production furnace campaign, are fused cast chrome-containing refractories, i. e., chrome-alumina-zirconia KhATs-30 and KhTs-45, and high-chrome refractories KhPL-85 and KhMG. However, use of pure raw materials in a charge for producing chrome-containing refractories of these grades is one of the reasons for their high cost. Therefore, involvement of different forms of industrial waste in the manufacture of **fused cast chrome-alumina-zirconia** and high-chrome refractories makes it possible to reduce their cost and to expand the volume of application[78].

提出了提高**熔铸刚玉耐火材料**生产质量和经济效益的主要技术指标和影响因素。通过观察 ZrO_2 ∶ SiO_2 = 2 ∶ 1 和 SiO_2 ∶ Na_2O≥12 的质量比，优化 BC-33 耐火材料的成分，可以得到具有高耐腐蚀性、玻璃相析出温度在 1450 ℃以上、熔融体中缺陷倾向最小的材料。只有使用耐火材料的氧化熔化，包括炉料的电弧熔化和熔体的气-气处理，才能达到这些质量指标[79]。

The main technological indicators and factors for improving the quality and economic efficiency of **fused-cast baddeleyite-corundum refractory** production are given. Materials with high corrosion resistance, glass-phase precipitation temperatures above 1450 ℃, and minimal tendency to precipitate defects in the molten mass could be produced by optimizing the BC-33 refractory composition while observing the mass ratios ZrO_2 ∶ SiO_2 = 2 ∶ 1 and SiO_2 ∶ Na_2O ≥12. Such quality indicators are achieved only if oxidative melting of the refractories is used, including arc-melting of the charge and gas-air treatment of the melt[79].

参 考 文 献

[1] An Jing, Xue Xiangxin. Life-cycle carbon footprint analysis of magnesia products [J]. Resources, Conservation and Recycling, 2017, 119: 4-11.

[2] Kuang J P, Harding R A, Campbell J. Investigation into refractories as crucible and mould materials for melting and casting γ-TiAl alloys [J]. Materials Science and Technology, 2000, 16 (9): 1007-1016.

[3] An Jing, Li Yingnan, Middleton R S. Reducing energy consumption and carbon emissions of magnesia refractory products: A life-cycle perspective [J]. Journal of Cleaner Production, 2018, 182: 363-371.

[4] Deng Zirong, Wei Yaowu, Zhou Hui, et al. Influences of solid content of slurry on the microstructure of CaO crucible prepared by slip casting [J]. International Journal of Applied Ceramic Technology, 2023, 20 (3): 1547-1556.

[5] Wang Juntao, Wang Qinghu, Chen Songlin. Comparison of corrosion resistance of high chromia brick, high zirconia brick, fused AZS brick and Si_3N_4 bonded SiC brick against coal slag [J]. Ceramics International, 2020, 46 (8): 10851-10860.

[6] Sun M O, Park J H. Corrosion behaviors of zirconia refractory by CaO-SiO_2-MgO-CaF_2 slag [J]. Journal of the American Ceramic Society, 2009, 92 (3): 717-723.

[7] Bennett J P, Kwong K S, Nakano J, et al. Impact of temperature and oxygen partial pressure on aluminum phosphate in high chrome oxide refractories [J]. 2014, 92: 248-257.

[8] Bouchetou M L, Poirier J, Arbelaez Morales L, et al. Synthesis of an innovative zirconia-mullite raw material sintered from andalusite and zircon precursors and an evaluation of its corrosion and thermal shock performance [J]. Ceramics International, 2019, 45 (10): 12832-12844.

[9] Chandra D, Das G, Maitra S. Comparison of the role of MgO and CaO additives on the microstructures of reaction-sintered zirconia-mullite composite [J]. International Journal of Applied Ceramic Technology, 2015, 12 (4): 771-782.

[10] Ebadzadeh T, Ghasemi E. Influence of starting materials on the reaction sintering of mullite-ZrO_2 composites [J]. 2000, 283 (1/2): 289-297.

[11] Kim H J, Kim J J, Lee J K. Enhancement of the surface roughness by powder spray coating on zirconia substrate [J]. Journal of Nanoscience and Nanotechnology, 2019, 19 (10): 6285-6290.

[12] Zhang X H, Wen D D, Deng Z H, et al. Study on the grinding behavior of laser-structured grinding in silicon nitride ceramic [J]. International Journal of Advanced Manufacturing Technology, 2018, 96 (9/10/11/12): 3081-3091.

[13] Hampshire S, Kennedy T. Silicon nitride-silicon carbide micro/nanocomposites: A review [J]. International Journal of Applied Ceramic Technology, 2022, 19 (2): 1107-1125.

[14] Kondo N, Kita K, Nagaoka T. Fabrication of silicon nitride from a slurry containing cellulose nanofibers [J]. Journal of the Ceramic Society of Japan, 2017, 125 (7): 588-590.

[15] Perevislov S N. Sintering behavior and properties of reaction-bonded silicon nitride [J]. Russian Journal of Applied Chemistry, 2021, 94 (2): 143-151.

[16] Perevislov S N. Structure, properties, and applications of graphite-like hexagonal boron nitride [J]. Refractories and Industrial Ceramics, 2019, 60 (3): 291-295.

[17] Liu Dong, Tang Chengchun, Xue Yanming, et al. New porous boron nitride materials [J]. Progress in Chemistry, 2013, 25 (7): 1113-1121.

[18] Li Baorang, Wen Bo, Chen Haozhi, et al. Corrosion behaviour and related mechanism of lithium vapour on aluminium nitride ceramic [J]. Corrosion Science, 2021, 178: 09058.

[19] Tanasta Z, Muhamad P, Kuwano N, et al. Reduction of defects on microstructure aluminium nitride using high temperature annealing heat treatment [C]. 3rd International Conference on Mechanical, Manufacturing and Process Plant Engineering (ICMMPE), 2017.

[20] Xue Zhang, Jin Du, Guo Shengsu, et al. Research progress on toughening and strengthening mechanism of ternary boride base cermets [J]. IOP Conference Series: Earth and Environmental Science, 2021, 692: 032068.

[21] Teker T, Sari M. Metallurgical properties of boride layers formed in pack boronized cementation steel [J]. Materials Testing, 2022, 64 (9): 1332-1339.

[22] Mani S, Palanisamy C, Murugesan M, et al. Estimation of distinctive mechanical properties of spark plasma sintered titanium-titanium boride composites through nano-indentation technique [J]. International Journal of Materials Research, 2015, 106 (11): 1182-1188.

[23] Grigoriev O N, Zhunkovski H L, Vedel D V, et al. Features of zirconium boride-chromium interaction [J]. Powder Metallurgy and Metal Ceramics, 2019, 58 (7/8): 455-462.

[24] Zhou Lijuan, Yin Kaili, Wang Tianqi, et al. Preparation and properties of ZrB_2 composite coatings by CVD [C]. 4th International Conference on Environmental Science and Material Application (ESMA), 2018, 252: 022101.

[25] Domi Y, Usui H, Sugimoto K, et al. Reaction behavior of a silicide electrode with lithium in an ionic-liquid electrolyte [J]. Acs Omega, 2020, 5 (35): 22631-22636.

[26] Kim J, Park Y C, Kumar M M D. Influence of temperature, metal layer, and groove angle in the nanowire growth: a prospective study on nickel silicide nanowires [J]. Journal of Nanoparticle Research, 2015, 17 (1): 15.

[27] Yao Z, Stiglich J, Sudarshan T S. Molybdenum silicide based materials and their properties [J]. Journal of Materials Engineering and Performance, 1999, 8 (3): 291-304.

[28] Yeo S C, Jeon S W, Lee J E, et al. Combustion characteristics of molybdenum-silicon mixtures [J]. Ceramics International, 2015, 41 (1): 1711-1723.

[29] Xu Jianguang, Zhang Baolin, Li Wenlan, et al. Synthesis of pure molybdenum disilicide by the " chemical oven" self-propagating combustion method [J]. Ceramics International, 2003, 29 (5): 543-546.

[30] Smirnov A N, Khobta A S, Smirnov E N, et al. Casting of steel from the tundish of a continuous caster with a sliding gate [J]. Russian Metallurgy (Metally), 2012, 12: 1048-1052.

[31] Tsukaguchi Y, Hayashi H, Kurimoto K, et al. Development of swirling-flow submerged entry nozzles for slab casting [J]. ISIJ International, 2010, 50 (5): 721-729.

[32] Real-Ramirez C A, Miranda-Tello R, Carvajal-Mariscal I, et al. Hydrodynamic study of a submerged entry nozzle with flow modifiers [J]. Metallurgical and Materials Transactions B, 2017, 48: 1358-1375.

[33] Yogeshwar S, Toshihiko E. Tundish technology for clean steel production [M]. World Scientific, 2007.

[34] Bhardwaj B P. Steel and iron handbook [M]. NewDelhi: National Institute of Industrial Research, 2014.

[35] Si Yaochen, Zhang Fan, Li Xin, et al. Thermodynamic calculation and microstructure characterization of spinel formation in $MgO\text{-}Al_2O_3\text{-}C$ refractories [J]. Ceramics International, 2022, 48 (11): 15525-15532.

[36] Wei Zheng, Liu Guoqi, Yang Jinsong. Effect of additives and carbon resources on thermal shock resistance of spinel-C [J]. Key Engineering Materials, 2017, 726: 440-444.

[37] Liu Guoqi, Li Hongxia, Yan Wengang, et al. Corrosion mechanism of functional refractories for continuous casting by free cutting steel [J]. Advanced Materials Research, 2010, 129-131: 348-352.

[38] Tang Haiyan, Guo Xiaochen, Wu Guanghui, et al. Effect of gas blown modes on mixing phenomena in a bottom stirring ladle with dual plugs [J]. ISIJ International, 2016, 56 (12): 2161-2170.

[39] Jauhiainen A, Jonsson L, Dongyuan S. Modelling of alloy mixing into steel [J]. Modelling of Alloy Mixing into Steel, 2001, 30 (4): 242-253.

[40] Madias J, Moreno A. Strategies against reoxidation of liquid steel in billet casting with metering nozzle [C] //Proceedings of the Iron & Steel Technology Conference. Warrendale: Association for Iron & Steel Technology, 2014, 1771-1778.

[41] Szmzrzsekera I V, Brimacombe J K. Evolution or revolution? —A new era in billet casting [J]. Canadian Metallurgical Quarterly, 1999, 38 (5): 347-362.

[42] Hossain SK S, Roy P K. Fabrication of sustainable insulation refractory: Utilization of different wastes [J]. Boletín de la Sociedad Española de Cerámica y Vidrio, 2019, 58 (3): 115-125.

[43] Hossain S S, Bae C J, Roy P K. A replacement of traditional insulation refractory brick by a waste-derived lightweight refractory castable [J]. International Journal of Applied Ceramic Technology, 2021, 18 (5): 1738-1791.

[44] Kremenetskaya I, Lvanova L, Chislov M, et al. Physicochemical transformation of expanded vermiculite after long-term use in hydroponics [J]. Applied Clay Science, 2020, 198: 105839.

[45] Mo K H, Lee Hongjie, Liu M Y J, et al. Incorporation of expanded vermiculite lightweight aggregate in cement mortar [J]. Construction and Building Materials, 2018, 179: 302-306.

[46] Lv Jianhua, Sun Bin, Jin Jing, et al. Mechanical and slow-released property of poly (acrylamide) hydrogel reinforced by diatomite [J]. Materials Science and Engineering: C, 2019, 99: 315-321.

[47] Qi Liqiang, Teng Fei, Liu Kunyang, et al. Effect of floating bead on the manufacturing process of fly ash ceramics [J]. International Journal of Applied Ceramic Technology, 2019, 17 (1): 122-129.

[48] Wang Qingping, Wang Hui, Min Fanfei, et al. Preparation and fire-resistant properties of fly ash floating beads/epoxy resin coatings [J]. Asian Journal of Chemistry, 2014, 26 (60): 1704-1706.

[49] Sang Guochen, Du Xiaoyun, Zhang Yangkai, et al. A novel composite for thermal energy storage from alumina hollow sphere/paraffin and alkali-activated slag [J]. Ceramics International, 2021, 47 (11): 15947-15957.

[50] Cai Jiabai, Li Huan, Jing Qi, et al. Embedding ruthenium nanoparticles in the shell layer of titanium zirconium oxide hollow spheres to catalyze the degradation of alkali lignin under mild condition [J]. Journal of Hazardous Materials, 2021, 411: 125161.

[51] Liu Jun, Shao Peng, Wang Shihao. The influence of diatomite on the strength and microstructure of portland cement [J]. MATEC Web of Conferences, 2016, 67: 07017.

[52] Al-Shmaisani S, Kalina R D, Ferron R D, et al. Assessment of blended coal source fly ashes and blended fly ashes [J]. Construction and Building Materials, 2022, 342: 127918.

[53] Sun Dan, Wang Lijiu. Utilization of paraffin/expanded perlite materials to improve mechanical and thermal properties of cement mortar [J]. Construction and Building Materials, 2015, 101 (1): 791-796.

[54] Jia Liao, Li Yuanbing, Li Shujing. Preparation of utralight alumina lightweight insulation brick [J]. 2014, 602-603: 648-651.

[55] Sutcu S, Akkurt S, Bayram A, et al. Production of anorthite refractory insulating firebrick from mixtures of clay and recycled paper waste with sawdust addition [J]. Ceramics International, 2012, 38 (2): 1033-1041.

[56] Sutcu S, Akkurt S. Utilization of recycled paper processing residues and clay of different sources

for the production of porous anorthite ceramics [J]. Journal of the European Ceramic Society, 2010, 30 (8): 1785-1793.

[57] Tuna Aydin. Development of porous lightweight clay bricks using a replication method [J]. Journal of the Australian Ceramic Society, 2018, 54: 169-175.

[58] Pye A. A review of asbestos substitute materials in industrial applications [J]. Journal of Hazardous Materials, 1979, 3 (2): 125-147.

[59] Ahmad R B, Mehrdad K, Mohammad H N F, et al. High temperature ablation of kaolinite layered silicate/phenolic resin/asbestos cloth nanocomposite [J]. Journal of Hazardous Materials, 2008, 150 (1): 136-145.

[60] Zhao Haiqian, Liu Xiaoyan, Liu Lijun, et al. Study on new thermal insulation construction of thermal recovery boiler [J]. Energy Procedia, 2012, 16 (C): 1466-1471.

[61] Goldsby G C, Yun H M, Morscher G N, et al. Annealing effects on creep of polycrystalline alumina-based fibers [J]. Materials Science and Engineering, 1988, 242 (1): 278-283.

[62] Xu Jianfeng, Fu Shunde, Zhang Ke, et al. Influence of heat treatment process on properties of polycrystalline alumina fiber [J]. Advanced Materials Research, 2011, 399-401: 822-827.

[63] Sun Jingjing, Hu Zijun, Li Junning, et al. Thermal and mechanical properties of fibrous zirconia ceramics with ultra-high porosity [J]. Ceramics International, 2014, 40 (8): 11787-11793.

[64] Rubing Zhang, Ye Changshou, Wang Baolin, et al. Novel Al_2O_3-SiO_2 aerogel/porous zirconia composite with ultra-low thermal conductivity [J]. Journal of Porous Materials, 2018, 25: 171-178.

[65] Ebert H P, Hemberger F. Intercomparison of thermal conductivity measurements on a calcium silicate insulation material [J]. International Journal of Thermal Sciences, 2011, 50 (10): 1838-1844.

[66] Zheng Qijun, Chung D O L. Microporous calcium silicate thermal insulator [J]. Microporous Calcium Silicate Thermal Insulator, 1990, 6 (7): 666-670.

[67] Guo Jingran, Fu Shubin, Deng Yuanpeng, et al. Hypocrystalline ceramic aerogels for thermal insulation at extreme conditions [J]. Nature, 2022, 606: 909-916.

[68] Xu Xiang, Zhang Qiangqiang, Hao Menglong, et al. Double-negative-index ceramic aerogels for thermal superinsulation [J]. Science, 2019, 363 (6428): 723-727.

[69] Hou Qingdong, Luo Xudong, Xie Zhipeng, et al. Preparation and characterization of microporous magnesia-based refractory [J]. International Journal of Applied Ceramic Technology, 2020, 17 (6): 2629-2937.

[70] Fu Lvping, Zou Yongshun, Huang Ao, et al. Corrosion mechanism of lightweight microporous alumina-based refractory by molten steel [J]. International Journal of Applied Ceramic Technology, 2019, 120 (6): 3075-3714.

[71] Wang Miao, Bai Hao, Zhao Lihua, et al. Research on the optimization application of BF hot stove energy-saving coating [C]. 5th International Congress on the Science and Technology of Ironmaking, Shanghai, China, 2011: 1007-1010.

[72] Han Rifei, Tariq N H, Zhao Feng, et al. High infrared emissivity energy-saving coatings based on $LaMnO_3$ perovskite ceramics [J]. Ceramics International, 2022, 48 (14): 20110-20115.

[73] Xue Zhijia, Shi Junrui, Wang Yue, et al. Research on the energy-saving property of the boiler far-infrared coatings [C]. 2011 International Conference on Materials for Renewable Energy & Environment. Shanghai, China, 2011: 782-785.

[74] Wang Runfeng, Gu Huazhi, Bai Chen, et al. Relationship between phase composition, thermal conductivity and corrosion resistance of fused-cast AZS refractories [J]. International Ceramic Review, 2021, 70: 56-61.

[75] Sengupta P, Mishra R K, Soudamini N, et al. Study on fused/cast AZS refractories for deployment in vitrification of radioactive waste effluents [J]. Journal of Nuclear Materials, 2015, 467: 144-154.

[76] Hell J, Vespa P, Cabodi I, et al. Blistering phenomenon of molten glass in contact with zirconia-based refractories [J]. Journal of the European Ceramic Society, 2021, 41 (10): 5359-5366.

[77] Yan Xinhua, Yuan Lei, Liu ZhenLi, et al. Preparation of porous mullite ceramic for high temperature flue gas filtration application by gel casting method [J]. Journal of the Australian Ceramic Society, 2021, 57 (4): 1189-1198.

[78] Sokolov V A, Gasparyan M D, Kirov S S. Preparation of fused cast chromium-corundum refractories using baddeleyite-corundum object scrap [J]. Refractories and Industrial Ceramics, 2013, 54 (4): 327-330.

[79] Sokolov V A, Gasparyan M D, Kiryukhin V V. Production technology features of fused-cast baddeleyite-corundum refractory [J]. Refractories and Industrial Ceramics, 2020, 60 (6): 543-547.

6 不定形耐火材料制品

6.1 耐火浇注料

6.1.1 术语词组

耐火浇注料 refractory castable
铝酸钙水泥结合浇注料 calcium aluminate cement-bonded castable
超低水泥耐火浇注料 ultra-low cement refractory castable
无水泥耐火浇注料 cement free refractory castable
磷酸盐结合浇注料 phosphate-bonded castable
可水合氧化铝结合浇注料 hydratable alumina bonded castable
铝-镁质浇注料 alumina-magnesia castable
氧化铝-碳化硅-碳质浇注料 Al_2O_3-SiC-C (ASC) castable
耐酸耐火浇注料 acid resistant refractory castable
耐碱耐火浇注料 alkaline resistant refractory castable
耐磨耐火浇注料 wearing resistant refractory castable
轻质（隔热）耐火浇注料 light weight refractory castable

6.1.2 术语句子

CAC 结合浇注料由于其具有良好的高温体积稳定性、高强度和抗热震性能，因此在炼钢生产中得到了广泛的应用。

CAC-bonded castables are widely used in the steel-making process because of their excellent high-temperature performances, such as good volume stability at high temperature, high thermal strength and thermal shock resistance.

高铝耐火浇注料作为一种重要的不定形耐火材料，在炼钢、水泥、石油化工等行业中有着广泛的应用。

High-alumina refractory castables, as one of the most important unshaped refractories, have been extensively used in the steel-making, cement, and petrochemical industries.

本工作研究了在**耐火浇注料**生产过程中，水合氧化铝在几种形式的氧化镁存在下的水化行为。

This work investigates hydration behavior of hydratable alumina in the presence of several forms of magnesia used for production of **refractory castables**.

低、超低、无水泥浇注料近年来得到了广泛的应用。

Low, ultralow and no cement castables have been widely used in recent years.

Al_2O_3-SiC-C（ASC）耐火浇注料因其良好的热机械性能、优异的抗氧化性和抗渣性而广泛应用于高炉槽道和渣道中。

Al_2O_3-SiC-C (ASC) refractory castables are widely used in blast furnace trough and slag runners, due to their favorable thermomechanical properties, excellent oxidation resistance and slag resistance.

以棕刚玉为骨料，碳化硅、白刚玉粉、氧化铝微粉、六偏磷酸钠等为辅料制备**氧化铝-碳化硅-碳质铁沟浇注料**。

Al_2O_3-SiC-C castables for iron runner were prepared with brown corundum as aggregates and with silicon carbide, white corundum powder, alumina micropowder, sodium hexametaphosphate, etc., as auxiliary materials.

6.1.3 术语段落

铝酸钙水泥（CAC）是浇注料主要的结合剂，通过水化胶结作用为浇注料提供脱模强度。养护温度是影响 CAC 水化速度、水化产物种类以及水化产物分布的重要因素，会对浇注料的性能产生很大影响。养护时间会影响 CAC 的水化程度和水化产物的分布，对浇注料性能产生一定的影响。本文主要探究了养护温度和养护时间对 **CAC 结合刚玉质浇注料**体积稳定性和力学性能的影响。研究了养护温度和养护时间对浇注料基质中 CAC 水化产物种类、水化程度、水化产物分布的影响，并分析了其对热处理过程中六铝酸钙（CA_6）生成温度和分布及对浇注料显微结构的变化的影响，建立了浇注料的体积稳定性以及力学强度与养护温度、养护时间、CAC 水化程度和水化产物分布的关系[1]。

Calcium aluminate cement (CAC) is an important binder, providing effective demolding strength via hydration process, for refractory castables. Curing temperature is an important factor affecting the hydration speed of CAC and the species of hydration products and the distribution of hydration products, which will have a great impact on the properties of castables. The curing time will also affect the hydration degree of CAC and the distribution of hydration products, which will have a certain impact on the properties of castables. This paper mainly studies the effects of curing temperature and curing time on the volume stability and mechanical properties of **CAC-bonded corundum-based castables**. The effects of curing temperature and curing time on the species of CAC hydration products, hydration degree and hydration products distribution in the castable matrix were studied. The formation temperature and distribution of calcium hexaluminate (CA_6) during firing and the effect of microstructure changes of the castable were analyzed. The relationship between volume stability and mechanical strength of castables and curing temperature, curing time, hydration degree of CAC and distribution of hydration products were established[1].

由于**刚玉-镁铝尖晶石基浇注料**具有优异的耐腐蚀性和抗热震性，在炼钢过程中被广泛用作钢包衬。目前，浇注料中的 $MgAl_2O_4$ 尖晶石成分可以通过添加合成尖晶石或通过 Al_2O_3-MgO 反应原位生成来引入。然而，合成尖晶石通常导致成本高且在浇注料中的分布不均匀。尖晶石的原位形成通常需要游离的 MgO，这会产生较大的体积膨胀，从而在高温下产生不稳定的物理性质。因此，近年来发展了含 $MgAl_2O_4$ 尖晶石的铝酸钙水泥。Wohrmeyer 合成了 CMA 水泥，并将其作为铝尖晶石和铝镁基浇注料的黏结剂。结果表明，CMA 显著提高了材料的耐蚀性和抗侵彻性。Khalil 发现，使用埃及白云石和活性氧化铝制备的 CMA 在强度和耐火度之间表现出折中。当在耐火级镁骨料中加入 10%的此类水泥时，耐火浇注体的热强度和抗热震性得到了改善[2]。

Due to their excellent corrosion and thermal shock resistance, **corundum-$MgAl_2O_4$ spinel-based castables** are widely used as ladle linings, purging plugs and well blocks in steel-making processes. Currently, the $MgAl_2O_4$ spinel constituent within castables can be introduced either through the addition of synthesized spinel or in-situ formation from a Al_2O_3-MgO reaction. However, the synthesized spinel typically leads to high cost and inhomogeneous distribution in the castables. The in-situ formation of spinel often requires free MgO, which produces large volume expansions and thus unstable physical properties at elevated temperatures. Therefore, in recent years, calcium aluminate cement with $MgAl_2O_4$ spinel (CMA) has been developed.

Wohrmeyer synthesized CMA cement and used it as a binder in alumina-spinel and alumina magnesia-based castables. The results showed that CMA significantly improved resistance to both corrosion and penetration. Khalil found that CMA prepared using Egyptian dolomite and reactive alumina exhibited a compromise between strength and higher refractoriness. When 10% of such cement was added to refractory-grade magnesia aggregate, the refractory castable bodies were shown to have improved hot-strength and thermal shock resistance[2].

本工作的主要贡献是证实了 CAC 水合物 CAH_{10}和 C_2AH_8向 C_3AH_6 和 AH_3 的转化不是通过固相反应发生的，而是通过溶液-沉淀过程发生的，并澄清了关于转化途径的长期争论。本工作为研究 **CAC 结合浇注料**中气孔通道的变化奠定了理论基础，也为提高浇注料的抗爆性能提供了新的研究路径[3]。

The main contribution of this work is the validation that the conversion of CAC hydrates CAH_{10} and C_2AH_8 to C_3AH_6 and AH_3 does not take place through the solid-state reaction, but happens through a solution-precipitation process and clarifying the long-running debate about the conversion approaches. This work lays the theoretical foundation for the investigation of the change of pore channel in **CAC bonded castables**, and also provides new research paths for improving the explosion resistance of castables[3].

可水合氧化铝（HA）是一种优良的无钙耐火结合剂，但其水化速度过快从而限制了**水合氧化铝结合浇注料**的使用时间。本研究采用旋转球磨机分别研磨 1 h和 6 h，研究研磨对水合氧化铝水化及水合氧化铝结合浇注料性能的影响。水合氧化铝样品在 30 ℃下养护，然后用冷冻真空干燥进行处理。研究了干燥后水合氧化铝样品的相组成和微观结构。此外，还对未经研磨以及研磨的水合氧化铝浇注料的流动性能和力学强度进行了研究。结果表明：由于水合氧化铝颗粒变小，从而堵塞了水合氧化铝颗粒表面的微孔和微裂纹，从而降低了水合氧化铝的水化率，提高了浇注料的流动能力，使得水合氧化铝颗粒比表面积减小[4]。

Hydratable alumina (HA) is a superior Ca-free refractory binder, but the quick hydration rate restricts the working time of **castables bonded with HA**. In this work, HA was grounded for 1 h and 6 h by a rotational ball mill to study the effect of grinding on the hydration of HA and properties of HA-bonded castables. HA samples with and without grinding were cured at 30 ℃ and then terminated by freeze-vacuum drying. The phase composition and microstructure of the dried HA samples were then examined. Moreover, flow ability and mechanical strength of castables containing ungrounded and grounded HA were also investigated. The results indicate that the specific area of HA particles were decreased by grinding as the micro-pores and micro-cracks on the surface of HA particles were blocked by smaller HA particles, thereby decreasing the hydration rate of HA and increasing the flow ability of castables[4].

Al_2O_3-SiC-C（ASC）耐火浇注料因其良好的热机械性能、优异的抗氧化性和抗渣性而广泛应用于高炉槽道和渣道中。尽管 ASC 浇注料在冶炼工业中得到了有效的应用，但在提高其性能和延长使用寿命方面仍面临着严峻的挑战。在高炉炼铁过程中，ASC 浇注料长期暴露在高温铁水和炉渣的腐蚀和冲刷中，并反复经受高温加热和冷却环境。这些导致了结构的持续变化、体积的收缩、熔渣的腐蚀和渗透，从而降低了 ASC 浇注料的使用寿命。因此，改善 ASC 浇注料的显微组织对稳定其高温性能、提高其使用寿命具有重要意义。为解决上述问题，开展了提高 ASC 浇注料显微组织、力学性能和抗渣性能的研究。例如，原位控制碳化硅和莫来石晶须的形成有利于孔隙结构的调整和裂纹的偏转，从而提高了耐火材料的高温断裂韧性。此外，SiC 晶须的形成和分布对 ASC 浇注料的抗热震性和抗渣蚀性有显著影响。Chen 等人的研究表明，红柱石聚集体的引入产生了富 SiO_2 的液相，导致二次莫来石的形成，从而提高了 ASC 浇注料的抗氧化性和热震稳定性。Fan 等有报道称，使用铝铬渣作为添加剂，β-Al_2O_3 与 SiO_2 反应形成液相，通过促进烧结和原位生成莫来石来提高材料的机械强度，从而提高 ASC 浇注料的性能。因此，原位形成 SiC 和莫来石晶须有利于提高 ASC 浇注料的力学性能、抗热震性和抗渣性[5]。

Al_2O_3-SiC-C (ASC) refractory castables are widely used in blast furnace trough and slag runners, due to their favorable thermomechanical properties, excellent oxidation resistance and slag resistance. Despite the effective use of ASC castables in smelting industry, there are still serious challenges in improving their performances and extending the service life. During blast furnace iron-making process, ASC castables were exposed to the corrosion and scouring of high-temperature molten iron and slag for a long time and were suffering from heating and cooling environment at high temperatures repeatedly. These led to continuous structural changes, volume shrinkage, corrosion and penetration of molten slag, and the resultant reduced service life of ASC castables. Therefore, it is of great importance to improve the microstructure of ASC castables to stabilize their high-temperature performances and elevate the service life. In order to resolve the above mentioned problems, researches have been carried out to promote the microstructure, mechanical properties and slag resistance of ASC castables. For example, the controlled in-situ formation of SiC and mullite whiskers was favorable of the adjustment of pore structure and crack deflection, resulting in the improved high temperature fracture toughness of refractories. Besides, the formation and distribution of SiC whiskers imposed significant influences on thermal shock resistance and slag corrosion resistance of ASC castables. Research by Chen et al suggested that introduction of andalusite aggregates produced SiO_2-rich liquid phase, leading to the formation of secondary mullite, and the resultant improved oxidation resistance and thermal shock stability of ASC castables. Fan et al reported the elevated properties of ASC castables by using alumina-chromium slag as additive, in which β-Al_2O_3 reacted with SiO_2 to form a liquid phase, improving the mechanical strength of materials from the promoted sintering and in-situ formation of mullite. Therefore, the in-situ formation of SiC and mullite whiskers is conducive to improve the mechanical properties, thermal shock resistance and slag resistance of ASC castables[5].

首次将活性氧化镁作为铝酸钙水泥（CAC）的替代水合黏结剂制备**含 Cr_2O_3 耐火浇注料**。研究了镁结合耐火浇注料热处理后 Cr(Ⅵ) 的形成及物理力学性能。通过对热处理后的 MgO 键合耐火浇注料基体的微观结构表征和相组成分析，全面了解了煅烧过程中 Cr(Ⅵ) 的抑制机制和强度发展。结果表明，经 700~1300 ℃下烧制后，活性氧化镁结合的浇注料中 Cr(Ⅵ) 的渗出量比铝酸钙水泥结合的浇注料低 6.7~28.1 倍。MgO 与 Cr_2O_3 和 Al_2O_3 的优先相互作用形成的原位 $Mg(Cr, Al)_2O_4$ 尖晶石是导致烧结过程中 Cr(Ⅵ) 的生成和强度发展受到抑制的主要原因[6]。

Reactive MgO was used in the first time as alternative hydraulic binder of calcium aluminate cement (CAC) to prepare **Cr_2O_3-bearing refractory castable**. The formation of Cr(Ⅵ), and physical and mechanical properties of MgO-bonded refractory castables after heat-treating were investigated. Microstructural characterization and phase composition analyses on the heat-treated MgO-bonded refractory castable matrices resulted in a comprehensive understanding of the mechanism for the inhibition of Cr(Ⅵ), and of the strength development during firing. The results indicate that compared with CAC, Cr(Ⅵ) levels were 6.7-28.1 times lower using reactive MgO after firing at 700-1300 ℃. The in situ $Mg(Cr, Al)_2O_4$ spinel formed from the preferential interactions among MgO and Cr_2O_3 and Al_2O_3 would be the main reason leading to the inhibited Cr(Ⅵ) formation and strength development during firing[6].

高铝耐火浇注料作为一种重要的不定形耐火材料，在炼钢、水泥、石油化工等行业中有着广泛的应用。与耐火砖相比，这种材料的主要优点包括：（1）安装方便；(2) 结构完整性好，没有任何接缝；(3) 不需要烧制过程，成本低，能耗低。有机或无机黏结剂在浇注料的生产中起着至关重要的作用，因为它们确保了浇注过程中混合物的可加工性和环境温度下的绿色机械强度。在各种黏结剂中，铝酸钙水泥 (CAC) 因其凝结时间适宜，粘接能力优异，是商用产品中最常用的一种。尽管有上述优点，但使用水泥的-·个显著缺点是，由于与浇注料中的 Al_2O_3 和 SiO_2 反应产生液相，在高温下会导致结构降解。为了解决这一问题，人们提出了各种新型黏结剂作为水泥的替代品 (如水化氧化铝、铝酸钙镁酸盐、胶体二氧化硅、亚微米氧化铝)，并系统研究了它们对水泥的微观结构、自流动值、力学性能、抗热震性和腐蚀行为的影响[7]。

High-alumina refractory castables, as one of the most important unshaped refractories, have been extensively used in the steel-making, cement, and petrochemical industries. The major advantages of this material compared to the refractory bricks include: (1) the ease of installation, (2) the excellent structural integrity without any joints, and (3) the low cost and energy consumption because the firing process is not needed. Organic or inorganic binders play a critical role in the production of the castables because they ensure the workability of the mixtures during casting and the green mechanical strength at ambient temperatures. Among the various binders, calcium aluminate cement (CAC) is the most commonly used one in commercial products because of its suitable setting time and excellent bonding ability. Despite the mentioned advantages, a significant disadvantage of using cement is the structural degradation at elevated temperatures owing to the generation of liquid phases by reacting with Al_2O_3 and SiO_2 in the castables. To solve this issue, a variety of new binders have been proposed as alternatives for cement (such as hydratable alumina, calcium magnesium aluminate, colloidal silica, submicron alumina), and their effects on the microstructure, self-flow values, mechanical properties, thermal shock resistance, and corrosion behavior have been systematically studied[7].

采用 SP-CA 水泥制备了**尖晶石-氧化铝浇注料**。经高温加热后，浇注料 RA12 和 RA13 的尖晶石尺寸为 0.1~0.4 μm，小于浇注料 RA14 (>0.5 μm) 的尖晶石尺寸。亚微米尖晶石颗粒直接结合在刚玉和碳钙镁石 (CA6) 晶粒上可以起到增强浇注料坯体的作用。在浇注料 AS 中，尖晶石尺寸较大的 (<88 μm) 并没有在尖晶石和氧化铝之间形成连接。此外，亚微米尖晶石可以减小浇注料基质的孔径[8]。

Spinel-alumina castables were fabricated with SP-CA cements. After heating at high temperature, the spinel sizes of castable RA12 and castable RA13 were 0.1-0.4 μm, which was smaller than that of castable RA14 (>0.5 μm). The submicron spinel particles directly bonded to corundum and hibonite (CA6) grains could strengthen the castable body. In castable AS, the large spinel size (< 88 μm) did not form linkages between spinel and alumina. Further, submicron spinel can reduce the pore size of the castable matrix[8].

6.2 耐火可塑料和耐火捣打料

6.2.1 术语词组

耐火可塑料 plastic refractory
刚玉耐磨可塑料 corundum wearable plastic refractory
黏土结合可塑料 clay bonded plastic refractory
磷酸盐结合可塑料 phosphate-bonded castable
捣打料 ramming refractory
铝-镁质捣打料 alumina-magnesia ramming refractory
高铝-碳化硅-碳质捣打料 Al_2O_3-SiC-C (ASC) ramming refractory
锆英石质耐火捣打料 zircon ramming refractory

6.2.2 术语句子

使用耐火耐酸砖、耐火混凝土、**耐火可塑料**和其他材料，用线标示或重新划线衬炉、窑炉、锅炉和类似装置。

Line or reline furnaces, kilns, boilers and similar installations using refractory or acid-resistant bricks, refractory concretes, **plastic refractories** and other materials.

高性能**耐火可塑料**的研制与应用。

Development and application of high performance **plastic refractory**.

结果表明：在 1000 ℃和 1300 ℃热处理后，随着蓝晶石添加量的增加，**耐火可塑料**的收缩率降低。

The results show that the shrinkage of **plastic refractory** decreased with increasing kyanite addition after heat-treated at 1000 ℃ and 1300 ℃.

自 2003 年以来，这种**捣打混合物**成功地用于炭黑反应器衬砌，确保了延长的使用寿命和高温服务。

Since 2003 year this **ramming mix** is successfully served in the carbon black reactors lining, ensuring the prolonged service period and high temperature service.

6.2.3 术语段落

根据 YB/T 5117—93《耐火黏土和**高铝耐火可塑料**线性变化试验方法》和 YB/T 5118—93《耐火黏土和高铝耐火可塑料强度试验方法》对烧结试样的永久线性变化、断裂模量和常温耐压强度进行了测试。热膨胀系数的测定按 GB/T 7320.1—2000《耐火材料：热膨胀推杆法的测定》。用游标卡尺测量收缩量，计算其永久线性变化和体积密度。断裂模量用 CT-1000 弯曲强度试验机（日本）测定，常温耐压强度采用 MS-20-S1 压缩试验机（日本）进行测试。采用 RPZ-03 热膨胀仪测定试样的热膨胀系数，采用 RZ-2A 高温抗热震炉对其抗热震性能进行了测试[9]。

The permanent linear change, modulus of rupture and clod crushing strength of fired specimens were tested according to YB/T 5117—93 (Test Method for Linear Change of Fireclay and **High Alumina Plastic Refractories**), and YB/T 5118—93 (Test Method for Strength of Fireclay and High Alumina Plastic Refractories). The thermal expansion coefficient was tested according to GB/T 7320.1—2000 (Refractory Materials: Determination of Thermal Expansion-push Rod Method). The shrinkage was measured using vernier caliper, and the permanent linear change and bulk density were calculated. The modulus of rupture was measured with CT-1000 bending strength testing machine (Japan). The clod crushing strength was measured with MS-20-S1 compression testing machine (Japan). The thermal expansion coefficient of speci-mens was measured with RPZ-03 thermal expansion instrument. The thermal shock resistance was measured with RZ-2A high temperature thermal shock resistance furnace[9].

本文模拟了一种主要应用于高温行业的具有高压实行为的**捣打混合料**，其具有高压实行为。这种材料具有吸收承受高热负荷的部件变形的能力。进行了三轴和仪表模具压实试验，以确定剪切和硬化行为，分别在 20～80 ℃的温度范围内进行捣打混合物的测试。当材料被压实时，特别观察到温度效应对材料响应的影响。夯实混合料性能的主要特征可以用改良剑桥模型的理论框架来表示。单个变量允许根据温度准确地再现硬化行为。此外，提出了在高压下的硬化行为的模型的扩展[10]。

This paper is devoted to the modelling of a specific **ramming mix** mainly used in the high-temperature industry due to its high-compacting behaviour. This material has the ability to absorb the deformation of parts submitted to high thermal loads. Triaxial and instrumented die compaction tests were carried out in order to identify the shear and hardening behaviours, respectively. Tests on the ramming mix were led for a temperature range between 20 ℃ and 80 ℃. The temperature effect is particularly observed on the material response when it is compacted. The main features of the behaviour of the ramming mix can be represented by the theoretical framework of the Modified Cam-Clay model. A single variable allows to accurately reproduce the hardening behaviour depending on the temperature. Moreover, an extension of the model for the hardening behaviour at high pressures is proposed[10].

为实现“双碳”目标，感应电炉以其高效节能的优势在铸造、冶金等行业越来越受到重视。感应电炉炉衬的使用寿命和安全性在很大程度上取决于炉衬，炉衬的膨胀和顺烧会导致炉衬的侵蚀和抗渣性。针对铝镁基干式捣打料的可剪裁性，以200目的新型多元铝酸钙镁（$CaO-MgO-Al_2O_3$，CMA）取代煅烧氧化镁颗粒，制备了**铝镁基干式捣打料**。对比评价了添加CMA的铝镁质干式捣打料的体积密度、显气孔率、强度和抗渣侵蚀性能。结果表明，锰渣在干捣打料中的针入度指数随CMA的加入先降低后略有增加。同时，干捣混合料的永久线性变化逐渐减小。当CMA的添加量（质量分数）达到4%时，干捣打混合料的强度略大于参考，并且渣渗透指数仅为后者的75%[11]。

To achieve the goal of " dual-carbon", induction furnaces with high efficiency and energysaving advantages are paid more attention in the foundry and metallurgy industries. The service life and safety of induction furnaces strongly depended on the lining because expansion and forward sintering could result in the erosion and slag resistance of the lining. Focusing on the tailoring properties of **alumina-magnesia-based dry ramming mixes**, calcined magnesia particles were replaced with the novel multi-component materials of calcium magnesium aluminate ($CaO-MgO-Al_2O_3$, CMA) with a size of 200 meshes. Properties such as the bulk density, apparent porosity, strength, and slag corrosion resistance of alumina-magnesia-based dry ramming mix containing CMA were evaluated contrastively. The results demonstrate that the penetration index of manganese-bearing slag in dry ramming mixes first decreased and then slightly increased with the addition of CMA. Meanwhile, the permanent linear change in dry ramming mixes was gradually reduced. When the addition (mass fraction) of CMA reached 4%, the strength of the dry ramming mixes was slightly greater than the reference, and the slag penetration index was just 75% of the latter[11].

以不同含量的 Al_2O_3、SiO_2 粉和 Al_2O_3 粉、$Al(H_2PO_4)_3/H_3PO_4$ 和铝酸钙水泥为黏结剂和混凝剂的四种铝土矿为原料，经混合、成型、固化、干燥等工艺制备了 **Al_2O_3-SiO_2 耐火可塑料**。通过在塑料耐火材料中引入不同品位的铝土矿，研究了不同 Al_2O_3 含量的铝土矿对塑料耐火材料 1100 ℃干燥和燃烧后基体体积密度、室温强度、线性变化率、耐磨性、相组成和微观结构的影响。结果表明，随着铝土矿中 Al_2O_3 含量的增加，Al_2O_3-SiO_2 塑料耐火材料的干燥烧结体积密度增大（这意味着铝土矿品位的提高），随之而来的是室温强度的降低。配方中开发的铝土矿品位将加强刚玉的结晶角色塑造。铝土矿和铝酸钙水泥中的杂质可以起到塑料耐火材料的助烧作用。铝土矿中 Al_2O_3 含量的增加会提高集料颗粒的强度，降低基体的烧结，减弱塑料耐火材料基体与集料的强度匹配，增加磨损量，降低耐磨性[12]。

Al_2O_3-SiO_2 plastic refractory was produced with raw materials including four kinds of bauxites with different content of Al_2O_3, SiO_2 powder and Al_2O_3 powder, $Al(H_2PO_4)_3/H_3PO_4$ and calcium aluminate cement as binding agent and coagulant, through the process of mixing, forming, curing, drying and so on. The effect of bauxite with different content of Al_2O_3 on bulk density, strength at room temperature, linear change rate, wear resistance, phase composition and microstructure in matrix of plastic refractories that were dried and burned at 1100 ℃ respectively was studied through the introduction of different grade of bauxite in the plastic refractories. The results show that the bulk density of Al_2O_3-SiO_2 plastic refractory dried and burned will increase with Al_2O_3 content in bauxite (meaning that the grade of bauxite was developed), which followed that the strength will decrease at room temperature. The crystalline characterization of corundum will be strengthened with bauxite grade being developed in the formula. The sintering aid of plastic refractories can be played by the impurities in bauxite and the calcium aluminate cement. The increasing Al_2O_3 in bauxite will increase the strength of aggregate particle, will reduce the sintering of the matrix, and will result that the strength match between the matrix and aggregate of plastic refractories will be weakened, which increases the amount of wear, and reduces the wear resistance[12].

磷酸盐结合耐火材料是广泛用于浇注料和**耐火可塑料**的一类重要材料。特别是**高铝磷酸盐结合耐火可塑料**，作为最重要的不定形耐火材料之一，与定形耐火材料相比，可以简单地获得并且容易修复，被广泛地用作各种工业锅炉和烟道衬里的修复材料。磷酸盐结合剂的种类很多，其中磷酸铝和磷酸铬铝在中高温（800~1400 ℃）下表现出良好的力学性能，导致它们广泛用于中温工作的锅炉中的耐火材料。然而，在该温度范围（800~1400 ℃）内，磷酸盐基耐火材料的强度也随着温度下降而降低。因此，有必要寻找一种提高中温锅炉用磷酸盐耐火材料使用寿命的方法[13]。

Phosphate-bonded refractory materials are an important class of materials widely used in castable and **plastic refractories**. Especially, **high-alumina phosphate-bonded plastic refractory materials**, which are one of the most important unshaped refractories and can be obtained simply and repaired easily compared to shaped refractories, are widely used in various industrial boilers and flue lining as repair materials. There are many kinds of phosphate binders, of which aluminum phosphate and chrome-alumina phosphate exhibit good mechanical properties at medium to high temperatures (800-1400 ℃), leading to their wide use in refractories in boilers working at medium temperatures. However, within this temperature range (800-1400 ℃), the strength of the phosphate-based refractory materials also decrease with falling temperature. Therefore, it is necessary to find an enhancement method to improve the service life of phosphate-based refractory materials that are used in medium-temperature boilers[13].

6.3 耐火喷射料

6.3.1 术语词组

喷射耐火材料 gunning refractory
喷补用耐火材料 gunned refractory
硅酸铝质喷射耐火材料 aluminium silicate gunning refractory
碱式喷射耐火材料 alkaline gunning refractory
高铝-碳化硅-碳质喷射耐火材料 Al_2O_3-SiC-C (ASC) gunning refractory
火焰喷补耐火材料 flame gunning refractory

6.3.2 术语句子

配制系列**硅酸铝质喷补耐火材料**并检测其抗热震性及喷补附着性能等。

A series of **aluminum silicate gunning refractories** and its properties, such as thermal shock resistance, adhesion strength, were measured.

配制系列**喷补用耐火材料**，并检测其烧后线变化及耐热震性等性能。

Gunning refractory are made and its properties, such as linear change during baking and thermal shock resistance, are measured.

本发明还提供在生产或修补冶金熔炉衬里中使用的**耐火喷补组合物**。

The invention also provides for **a refractory gunning composition** for use in the production or repair of metallurgical furnace linings.

喷射耐火材料具有耐火度高、机械强度好、耐磨性好等优异性能，已被广泛应用于冶金、水泥、化工、石油等众多领域的工业炉炉衬材料。

Spray refractories have been widely used as lining materials for industrial furnaces in numerous fields, such as metallurgy, cement, chemicals, and petroleum, and have excellent properties including high refractoriness, good mechanical strength, and good wear resistance.

对喷射成型的耐火材料产品来说，堆积比重与**喷射成型耐火材料产品**的致密度有密切关系。

For spray formed refractory products, the packing density is closely related to the density of **spray formed refractory products**.

等离子喷涂涂层使用工业废料，即铝灰渣制备，以研究其中的 Al_2O_3 和 $MgAl_2O_4$ 是否可以保留在涂层中，以及等离子喷涂工艺能否消除灰渣中始终存在的 AlN 相。

Plasma sprayed coatings were synthesized from using industrial waste, i. e. Aluminium dross to explore if the structural phases Al_2O_3 and $MgAl_2O_4$ in the dross could be retained in the coatings and also if the plasma spray process could eliminate the AlN phase always present in the dross.

纵览近年来不定形耐火材料的技术进展可以看出，**喷射耐火材料**技术进步较快，包括喷射工艺与喷射装备，以及喷射材料的作业性的控制技术。

Looking at the technical progress of amorphous refractory materials in recent years, it can be seen that the technology of **sprayed refractory materials** has made rapid progress, including spraying process and spraying equipment, as well as the control technology of the workability of sprayed materials.

6.3.3 术语段落

Wei 等研究了硅溶胶对**高铝低水泥喷涂耐火材料**性能的影响，发现添加 3% 的硅溶胶后，纳米 SiO_2 吸附在活性 α-Al_2O_3 颗粒表面，填补了颗粒之间的空隙。通过采取适当的工艺措施，莫来石的形成可以提高耐火材料的综合性能。然而，迄今为止，COREX 尚未对高铝质喷射耐火材料的抗熔渣侵蚀性能进行研究。在 MG 运行过程中，高温熔融灰渣中的 FeO、TiO_2、K_2O、Na_2O 等氧化杂质由于气流速度较快，可通过气孔或微裂纹穿透耐火材料。由于化学反应，产生了低熔点相；然而，耐火材料的物理性能和显微结构会恶化。此外，在现有 MGs 下，喷涂耐火材料在服役期间剥落严重[14]。

Wei et al. elucidated the effects of silica sol on the properties of **the high-alumina low-cement spray refractory** and found that upon adding 3% silica sol, nano-SiO_2 adsorbs on the active α-Al_2O_3 particle surfaces and fills the gap between the particles. By adopting appropriate technological measures, mullite formation can improve the comprehensive performance of the refractory. However, the resistance of high-alumina spray refractory to corrosion by molten slag has not been investigated in COREX so far. During the operation of the MG, FeO, TiO_2, K_2O, Na_2O, and other oxidation impurities in the high-temperature molten ash slag can penetrate the refractory through pores or micro-cracks owing to the highspeed of gas flow. A low-melting-temperature phase is generated because of the chemical reactions; however, the physical properties and microstructure of the refractory deteriorate. Moreover, with the current MGs, the spray refractory peels severely during its service period[14].

所得结果清楚地表明，疲劳裂纹可在**热喷涂涂层**中扩展，其扩展速率可用常用于块体材料的 Paris 公式描述。未检测到近阈值行为，即裂纹扩展速率随 K_{max} 减小而急剧减小。疲劳裂纹与静态断裂的断口形貌十分相似，仅能发现极少数具有条纹状表面形貌的穿晶断裂或断口烧蚀等疲劳损伤的局部断口形貌特征[15]。

The obtained results clearly indicate that fatigue crack can propagate in **thermally sprayed coatings** and the rate of propagation can be described by Paris law commonly used for bulk materials. The near-threshold behavior, i. e. , sharp decrease in crack growth rate with decreasing K_{max}, was not detected. Fracture morphology of fatigue crack and static rupture are very similar, and only few localized fractographic sign of fatigue damage such as transgranular fracture with striation-like surface morphology or fracture ablation can be found[15].

研究**喷补料**的流变性能是优化修补层质量和喷补作业的有效途径。定量评价了耐火材料的流动性以及耐火材料颗粒-水悬浮液的流变行为。研究了耐火材料颗粒-水悬浮液流动性与喷补料流动性的关系，以及喷补料流动性与附着能力的关系。实验采用特殊的流变仪和喷枪机进行。结果表明，某些参数对喷补料的流变行为和喷补过程有较强的影响。讨论了在这种悬浮液中导致凝胶结构形成的粒子和水之间的相互作用[16]。

Study on rheological properties of **gunned refractory** is an effective way to optimize the quality of repaired layer and gunning operation. The fluidity of refractory was quantitatively evaluated as well as the rheological behavior of refractory particle-water suspensions. The relationship between the fluidity of refractory particle-water suspension and that of gunned refractory, and the relationship between the fluidity and adhesion ability of gunned refractory were studied. The experiments were carried out with a special rheometer and gunning machine. The results show that there is strong influence of some parameters on the rheological behavior of gunned refractory and gunning process. Interactions between particles and water that lead to the formation of gelatinous structures in this suspension are discussed[16].

以铝矾土为原料，铝酸钙水泥为结合剂，研究了不同蓝晶石添加量对**铝矾土基喷补料**性能的影响。结果表明，在 1300 ℃和 1500 ℃热处理后，铝矾土基喷补料的线收缩率随着蓝晶石加入量的增加而降低。1300 ℃热处理后，铝矾土基喷补料的抗折强度随着蓝晶石添加量的增加先增大后减小；铝矾土基喷补料经 1500 ℃热处理后的抗折强度随蓝晶石加入量的增加而降低。当蓝晶石添加量（质量分数）为 5%时，铝矾土基喷补料的抗压强度降低。当蓝晶石添加量为 10%时，铝矾土基喷补料的抗压强度增加。在本实验中，蓝晶石的最佳添加量为 10%[17]。

The effects of different additions of kyanite on properties of **bauxite-based gunning refractory** were investigated with bauxite as raw materials and calcium aluminate cement as binders. The results showed that the linear shrinkages of the bauxite-based gunning refractory decreased with the increasing of kyanite addition after heat treatment under 1300 ℃ and 1500 ℃. The exural strength of the bauxite-based gunning refractory increased at first and then decreased with the increasing of kyanite addition after heat treatment under 1300 ℃; The exural strength of the bauxite-based gunning refractory decreased with the increasing of kyanite addition after heat treatment under 1500 ℃. The compression strength of the bauxite-based gunning refractory decreased when the kyanite addition (mass fraction) was 5%. The compression strength of the bauxite-based gunning refractory increased when the kyanite addition was 10%. In this experimemt, the best addition of kyanite was 10%[17].

与压入料在线造衬可节能一样，耐火浇注料湿式喷射施工技术近年越来越受到人们的关注。湿式喷射浇注料在低加水量条件下仍具有一定的自流性，且不易出现颗粒偏析现象，这不同于常规浇注料；**湿式喷射浇注料**的黏滞阻力低，易于管道输送，这不同于自流浇注料。湿式喷射浇注料还有一个更显著的特点就是喷射施工时需添加促凝剂，在促凝剂的作用下，使喷射到施工面上的浇注料迅速失去流动性而附着，可大大降低材料的回弹率[18]。

Same as online lining with pressed material, wet shot creting technology of refractory castable has been paid more and more attention in recent years. The wet spray castable still has a certain degree of self-flowing under the condition of low water addition, and it is not easy to appear particle segregation, which is different from the conventional castable. The **wet jet castable** has low viscous resistance and is easy to pipeline, which is different from the self-flow castable. Another more significant feature of wet jetting castables is that coagulants need to be added during jetting construction. Under the action of coagulants, the castables sprayed on the construction surface quickly lose fluidity and attach, which can greatly reduce the rebound rate of the material[18].

氧化镁质喷补材料因其优异的抗碱性（高 CaO/SiO_2 比）渣侵蚀性能而被广泛应用于钢铁制造过程中的工作衬里。磷酸盐是镁质喷补料中最常用的黏结剂，然而磷酸盐的使用增加了钢液中的磷含量。为了避免这种情况，胶体二氧化硅常被用作含氧化镁应用的结合剂。硅溶胶结合镁质喷补料可以安全、快速地烘干，减少了整体加工时间，降低了爆裂的风险。与其他喷补料相比，硅溶胶结合镁质喷补料还具有良好的烧结性和体积稳定性。然而，硅溶胶结合镁质喷补材料中镁砂的水化仍是一个需要解决的问题[19]。

Magnesia gunning materials are widely employed in the working lining during steel manufacturing due to their excellent corrosion resistance to basic (high CaO/SiO_2 ratio) slag. Phosphates are the most commonly used binders in magnesia gunning materials, however, their use increases the phosphorus content in molten steel. To avoid this, colloidal silica is often used as the binding agent for magnesia containing applications. Colloidal silica-bonded magnesia gunning materials can be safely and quickly dried, reducing the overall processing time and the risk of explosive spalling. Colloidal silicabonde magnesia gunning materials also have good sinterability and volumetric stability compared with other gunning materials. However, the hydration of magnesia in colloidal silica-bonded magnesia gunning materials is still a concern that needs to be addressed[19].

喷射混凝土是通过将骨料、胶凝材料和水的混合物喷射到隧道开挖中的岩石表面等接收基底上而产生的。根据地区和施工特点，喷射混凝土可应用于干混或湿混工艺。对于干混喷射混凝土，在喷头处的干混料（即水泥、集料、外加剂和固体掺合料）中加入水。对于湿喷混凝土，生产预拌混凝土（水、水泥、集料、外加剂和掺合料）并将其泵送到喷嘴，在喷嘴处与凝结速凝剂混合，然后喷射加压空气。由于速凝剂和/或特殊黏结剂的使用以及喷射工艺本身的原因，喷射混凝土的化学和结构性能与现浇混凝土存在显著差异。此外，为了提高喷射混凝土的耐久性能和减少碳足迹等环境影响，喷射混凝土用水泥正越来越多地被辅助胶凝材料（SCMs）替代[20]。

Shotcrete is produced by spraying a mixture of aggregates, cementitious materials and water onto a receiving substrate such as a rock face in tunneling. Depending on regional and construction specifics, the shotcrete can either be applied in the dry-mix or the wet-mix process. For dry-mix shotcrete, water is added to a dry mix (i. e. cement, aggregates, additions and solid admixtures) at the spraying nozzle. For wet-mix shotcrete, ready-mix concrete (water, cement, aggregates, additions and admixtures) is produced and pumped to the nozzle where it gets intermixed with the setting accelerator and consequently sprayed with pressurised air. Due to the use of accelerator and/or special binders and the spraying process itself, the chemical and structural properties of shotcrete markedly differ from those of cast concrete. Additionally, the cement used for shotcrete is being increasingly substituted with supplementary cementitious materials (SCMs) to improve the durability performance and to decrease the environmental impact such as carbon footprint[20].

6.4 干式耐火振捣料

6.4.1 术语词组

干式耐火振捣料 dry vibratable refractory
硅质干式振捣料 silica dry vibratable refractory
硅酸铝质干式振捣料 aluminium silicate dry vibratable refractory

刚玉质干式振捣料 corundum dry vibratable refractory

镁质干式振捣料 magnesia dry vibratable refractory

感应炉用碱性干式振捣料 alkaline dry vibratable refractory for induction fuenace

中间包用碱性干式振捣料 alkaline dry vibratable refractory for tundish

电炉底用镁钙铁质干式振捣料 alkaline dry vibratable refractory for electric furnace bottom

6.4.2 术语句子

目前使用的中间包**干式振捣料**多为镁质、铝质、铝硅质等。镁钙质中间包干式振动料具有抵抗熔渣侵蚀能力强、对钢水具有一定的洁净能力等优点。

At present, most of the intermediate ladle **dry vibrating materials** used are magnesium, aluminium and aluminium-silicon. Magnesium and calcium intermediate ladle dry vibrating material has the advantages of strong resistance to slag erosion and a certain clean ability to the steel.

MgO-Al_2O_3 质干式振捣料及其配套的封口料、修平料、隔离料在 700 kW 感应器上使用寿命在半年以上，而在其他感应器上使用寿命达 1~2 年。

MgO-Al_2O_3 dry vibrating materials and their matching sealing, levelling and isolating materials have a service life of more than half a year on 700 kW inductors and 1 to 2 years on other inductors.

以中档镁砂和电熔镁砂为主要原料，采用偏硅酸盐取代酚醛树脂为结合剂，并引入一定量的中温和低温增强剂，重点研究了结合剂的选择、颗粒级配的优化及外加物的引入等因素对中间包**碱性无碳干式振捣料**性能的影响，并探讨偏硅酸盐结合碱性无碳干式振动料的结合机理。

The effect of the choice of binding agent, the optimization of particle gradation and the introduction of admixtures on the performance of alkaline carbon free dry vibrating materials in the middle pack was investigated, and the bonding mechanism of **alkaline carbon free dry vibrating materials** combined with metasilicate was discussed.

通过在中间包**干式料**布料完毕振动前增加人工捣料，显著增加干式料致密度和强度，有效延长中间包使用寿命。

By adding manual pounding to the **dry material** of the intermediate ladle before it is finished vibrating, the density and strength of the dry material are significantly increased, effectively extending the service life of the intermediate ladle.

氧化铬粉的加入可赋予**干式振捣料**良好的抗侵蚀性能，脱硅锆与铁铝尖晶石粉的加入有助于改善干式振捣料的抗热震性能。

The addition of chromium oxide powder gives the **dry vibrating compound** good resistance to erosion and the addition of desilicon zirconium and iron aluminium spinel powder helps to improve the thermal shock resistance of the dry vibrating compound.

6.4.3 术语段落

为了优化 **MgO 基干式振捣料（DVM）** 的微观结构和热力学性能，引入金属 Fe 作为一种新型烧结助剂及其对物理和力学性能、负荷下的耐火度以及蠕变的影响行为进行了调查。传统的 MgO 基 DVM 存在液相烧结机制和有限的固固结合，导致推荐的工作温度较低。因此，致密化和直接键合都得到了极大的改善。此外，还获得了优异的热机械性能，与传统方法相比，$T_{0.5}$提高了 100 ℃，静止阶段蠕变速率降低了 90%。认为液相的键合类型、孔隙率和分布的变化是长期结构稳定性的主要机制[21]。

With the aim to optimize the microstructure and thermo-mechanical properties of **MgO-based dry vibratable material (DVM)**, metal Fe was introduced as a novel sintering aid and its effect on the physical and mechanical properties, refractoriness under load, as well as creep behavior was investigated. A liquid phase sintering mechanism and limited solid-solid bonding were identified for the traditional MgO-based DVM, which resulted in a low recommended working temperature. As a consequence, both densification and direct bonding were greatly improved. Moreover, excellent thermo-mechanical properties were also obtained, achieving an increase of 100 ℃ in $T_{0.5}$ and a decrease of 90% in creep rate at stationary stage compared to the traditional one. The changes in bonding type, porosity, and distribution of the liquid phases are considered as the main mechanisms for the long-term structural stability[21].

由于采用高纯原料，加入特殊的结合剂及外加剂，所研制的**干式料**完全能够满足高效长寿连铸化浇钢的要求，使用寿命大幅度提高。与镁质涂抹料相比，应用该中间包干式料可使中包钢壳的温度降低，这对改善操作环境、减轻劳动强度、降低钢水热量损失及防止钢壳蠕变都具有良好的作用，还可以减少钢水夹杂，净化钢水。以固体粉状酚醛作结合剂的干式料中加入适量的促烧剂、外加剂等可获得较好的使用性能。与镁质涂料中包工作衬相比，中间包的工作时间可达 48 h 以上。试验用中间包干式料使用完毕后，在靠近永久层端有 20 mm 松散层，材料呈现本身的性能，这对后续翻去工作层非常有利，对中间包永久层浇注料及钢壳都有一定的保护作用[22]。

Due to the use of high purity raw materials, the addition of special binding agents and additives, the developed **dry material** can fully meet the requirements of efficient and long life continuous casting of cast steel, the service life is greatly improved. Compared with the magnesium coating material, the application of this intermediate ladle dry material can make the temperature of the middle ladle steel shell lower which has a good effect on improving the operating environment, reducing the labour intensity, reducing the heat loss of the steel and preventing creep of the steel shell, but also reducing the steel inclusions and purifying the steel. With solid powder phenolic as a binding agent in the dry material, the addition of the appropriate amount of accelerator, additives, etc., can obtain better performance. Compared with the magnesium coating in the package work lining, the working time of the intermediate package up to 48 h or more. After the test intermediate ladle dry material is used, there is a 20 mm loose layer near the permanent layer end, the material presents its own properties, which is very favourable for the subsequent turning over of the working layer, and has a certain protective effect on the intermediate ladle permanent layer casting material and steel shell[22].

干式振捣料是不加任何液体黏结剂的干式耐火混合料，经振动获得致密施工体，无须养护，只经烘烤即能直接使用的不定形耐火材料，其工艺原理是：使用具有热固性有机及无机复合结合剂，粉料和骨料采用电熔和高纯氧化镁精心配制物料的颗粒级别，保证在较小的振动力作用下，颗粒能迅速移动，填充各种空隙而各级料无剩余和不足，从而获得较高的堆积密度，经烘烤便可硬化成型。工艺要求：（1）具有良好的施工性能，颗粒度、配比适宜，振动时，能快速形成一定的堆积密度。（2）具有较高的耐火度、强度及良好的耐钢水、熔渣的侵蚀性能。（3）在烘烤和浇钢过程中，不与永久层分离，有良好的整体性和附着力，不剥落不开裂。（4）对钢水质量无不良影响。（5）浇钢结束后，工作层残衬能自行解体，并与永久层分离[23]。

The **dry pounding material** is a dry refractory mixture without any liquid binder, obtained by vibration dense construction body, without maintenance only by baking that can be used directly in the indefinite refractory material, its process principle is: the use of thermosetting organic and inorganic composite bonding agent, powder and aggregate using electrofusion and high-purity magnesium oxide carefully formulated material particle level, to ensure that under the action of a small vibration force, the particles can quickly move. The material can be filled with various voids without residual or deficient material at all levels, thus obtaining a high bulk density, which can be hardened and formed by baking. Process requirements: (1) Has good construction performance, granularity, suitable ratio, vibration, can quickly form a certain density of accumulation. (2) High refractoriness, strength and good resistance to steel and slag erosion. (3) In the process of baking and pouring steel, it does not separate from the permanent layer, has good integrity and adhesion, does not flake and does not crack. (4) No adverse effect on the quality of the steel. (5) After pouring steel, the residual lining of the working layer can disintegrate on its own and separate from the permanent layer[23].

镁质干式料耐火度、荷重软化温度高，具有优异的抗碱性渣侵蚀性能。但**镁质干式料**的抗热震性能差，烧结温度比较高，由于炉子间断作业频繁，炉衬材料容易因热冲击而出现龟裂和剥落，因此，碱性干式料适用于小型容器。**镁钙质干式料**具有高熔点、抗碱性渣侵蚀和洁净钢水等优良性能，通常用在高度洁净的钢水部位以避免钢水污染，提高钢水洁净度，镁钙质干式料被广泛应用于中间包。氧化铁是镁钙质的烧结剂，通过相图分析，在加入氧化铁时，材料对 SiO_2 比较敏感，SiO_2 含量（质量分数）应该小于 1%[24]。

Magnesian dry charge has high refractoriness and load softening temperature and has excellent resistance to alkaline slag erosion. However, the **magnesia dry type material** has poor thermal shock resistance, the sintering temperature is relatively high, due to the frequent intermittent operation of the furnace, the furnace lining material is prone to cracking and flaking due to thermal shock, therefore, the alkaline dry type material is suitable for small containers. **Magnesia calcium dry material** has high melting point, resistance to alkaline slag erosion and clean steel and other excellent properties, usually used in the highly clean steel parts to avoid steel pollution, improve steel cleanliness, magnesia calcium dry material is widely used in the intermediate ladle. Iron oxide is the sintering agent for magnesium and calcium, through phase diagram analysis, when adding iron oxide, the material is more sensitive to SiO_2, the SiO_2 content (mass fraction) should be less than 1%[24].

6.5 耐火挤压料与压注料

6.5.1 术语词组

耐火挤压料 refractory mud
压注料 injection mix
Al_2O_3-SiO_2-SiC-C 质炮泥 Al_2O_3-SiO_2-SiC-C mud
有水炮泥 taphole mud

无水炮泥 waterless mud | 耐火压注料 injection refractory

6.5.2 术语句子

重点介绍用后钢包浇注料的回收工艺及回收物料在**高炉出铁口炮泥**中的应用。

The process of recovery of refractories for ladle and the application of recovery material in **blast furnace tap hole clay** are introduced emphatically.

有水炮泥的添加剂中包括膨胀剂、润滑剂和助烧结剂。

The additives of **water gun clay** include swelling agent, lubricant and sintering agent.

攀钢高炉炉前开铁口设备一直使用简易的电动开铁口机，已不适应高炉高强度冶炼和铁口新型**炮泥**的要求。

Pansteel blast furnace used simply electric taphole drill for a long time, which cannot apply to higher smelting intensity and the new **gun clay**.

6.5.3 术语段落

耐火挤压料是指采用液压式挤压机（俗称泥炮或泥枪）施工的耐火材料，是一类半硬质塑性耐火泥料，这类泥料主要用于堵塞高炉出铁口，俗称炮泥。而耐火压注料是指采用泥浆泵压注施工的耐火材料，是一类具有自流性能的耐火泥料，与自流浇注料作业性相似。这类材料主要用于充填耐火制品砌体与砌体之间，或耐火制品砌体与炉壳之间的缝隙[25]。

Refractory extrusion material refers to the construction of refractory materials using hydraulic extruder (commonly known as mud gun), is a class of semi-hard plastic refractory clay, this kind of clay is mainly used to block the blast furnace outlet, commonly known as mud. The refractory injection material refers to the refractory material which is constructed by pressure injection of mud pump, and it is a kind of refractory mud with self-flowing performance, similar to the self-flowing castable operation. These materials are mainly used to fill the gap between refractory masonry and masonry, or between refractory masonry and furnace shell[25].

炮泥耐火材料是功能性耐火材料，可用于保护高炉壁。炮泥的特性对炼铁高炉的安全性和高效率非常重要。炮泥耐火材料传统上是用刚玉、黏土、Fe-Si 的原料制备的 Fe-Si_3N_4、铝土矿砂、焦油和焦炭。原材料成本高（刚玉和 Fe-Si_3N_4），抑制了炮泥耐火材料的大规模应用。随着钢铁行业进入衰退期，降低耐火材料成本对炼钢厂具有重要意义[26]。

Taphole clay refractories are functional refractory materials which can be used to protect blast furnace wall. The properties of taphole clay are important to the safety and high efficiency of ironmaking blast furnace. Taphole clay refractories are traditionally prepared with raw materials of corundum, clay, Fe-Si_3N_4, chamotte, tar and coke. The high cost of raw materials (corundum and Fe-Si_3N_4) restrains the large-scale applications of taphole clay refractories. Lowering the cost of refractories is of great significance to the steel making plant as the iron and steel industry enters its recession period[26].

原始的**出水口泡泥耐火材料**采自中国一家钢铁厂。为了研究其性能，将不同含量的合成 Fe-Sialon-Ti(C, N) 复合材料添加到原始的出水口泡泥耐火材料中，添加的复合材料含量（质量分数）分别为 0%、5%、10%、15%和 20%。使用 100 MPa 的压力将样品压制成尺寸为 150 mm×25 mm×25 mm 的定形制品，然后在还原气氛下加热至 1400 ℃进行强度测试。此外，还制备了用于抗渣性测试的坩埚样品，其尺寸为 ϕ50 mm×50 mm，内部有一个孔（ϕ20 mm×25 mm）。在这些坩埚样品中加入了 10 g 高炉渣。随后对样品的体积密度、显气孔率和抗弯强度进行了测试。在 1500 ℃还原气氛中进行 3 h 抗渣试验。抗渣测试后，沿中心轴切割样品以检验抗渣性[26]。

The original **taphole clay refractories** were collected from a steel plant in China. Different contents of the as-synthesized Fe-Sialon-Ti(C, N) composites were added into original taphole clay refractories. The content (mass fraction) of Fe-Sialo-Ti(C, N) composites added to taphole clay refractories were 0%, 5%, 10%, 15% and 20%. The samples were shaped with the size of 150 mm×25 mm×25 mm under the pressure of 100 MPa for strength tests. Crucible samples for slag resistance test were shaped with the size of ϕ50 mm×50 mm and an inner hole (ϕ20 mm×25 mm). All samples were heated at 1400 ℃ in the reducing atmosphere. Then the bulk density, apparent porosity, and bending strength of samples were tested. Blast furnace slag (10 g) was added in crucible samples. Slag resistance test was carried out at 1500 ℃ in the reducing atmosphere for 3h. After the slag resistance test, crucible samples were cut along the central axes to examine slag resistance[26].

6.6 耐火泥浆和耐火涂料

6.6.1 术语词组

耐火泥浆 refractory mortar
耐火涂料（涂抹料）refractory coating
特种耐火泥浆 special refractory mortar
气硬性耐火泥浆 air setting jointing material; refractory mortar of air setting
水硬性耐火泥浆 refractory mortar of hydraulic setting
热硬性耐火泥浆 refractory mortar of heat setting
硅酸铝质隔热耐火泥浆 alumina-silica insulating refractory mortars
热辐射涂料 heat radiative coating
耐酸耐火涂料 acid resistant refractory coating
耐碱耐火涂料 alkaline resistant refractory coating

6.6.2 术语句子

分析了加热炉水冷梁外层耐火材料损毁机理，研制了新型**耐火喷涂料**，确定了水冷梁用半干法喷涂料的配方。

The damage mechanism of outer refractory in water-cooling beam of heating furnace is analysed. The new **spraying coating** is researched and the formulation of spraying coating is confirmed.

耐火泥浆是用来砌筑耐火砖作为工作衬和非工作衬，起着粘接耐火砖的作用。

Refractory mortar is used to build refractory bricks as working lining and non-working lining, which plays the role of bonding refractory bricks.

耐火泥浆力学性能的研究主要集中在获取材料的拉伸、剪切和压缩性能。

Studies of mechanical properties of **refractory mortars** have predominantly concentrated on the acquiring of global material properties in tension, shear and compression.

陶瓷涂层的重要作用是在铸件的铸造、凝固和成型过程中，在砂型基体和液态金属流动之间形成一个有效的耐火屏障。

The vital role of **earthenware coatings** is to structure a competent refractory barrier between the sand substrate and liquid metal flow during the period of casting, solidification and forming of the castings.

据亚伦称，金属铝可应用于耐火材料，在烧制后形成**耐火涂层**，保护耐火材料免受熔渣腐蚀。

According to Aram, metallic aluminum may be applied to refractories, which after firing forms a **refractory coating** that offers protection of the refractory against the corrosion of slags.

6.6.3 术语段落

所进行的分析表明，调查中的**耐火泥浆**的失效是由一系列离散的局部事件引起的。在剪切过程中，这些事件既可以是晶粒滑移，也可以是晶粒拉伸脱键。在整体剪切应力-应变曲线上，破坏事件由不同量级的峰值标记。在压缩作用下，由于颗粒及其团簇的重新排列，产生裂纹和孔隙闭合。这产生了指数应力-应变曲线，其中刚度随载荷增加。泥浆孔隙率既能降低整体破坏应力，又能促进止裂。黏结剂通过改变晶粒间的结合和摩擦来影响晶粒的黏聚力。摩擦力的影响在侧向力存在时更为明显。在晶粒间有更高摩擦的材料更硬，更强，有更少的脆性后破坏行为。未成熟黏结剂泥浆的特点是裂缝分叉强烈。这与孔隙率的影响相结合，导致高压缩性和低强度以及高止裂能力。成熟黏结剂的裂纹分支是由晶粒的黏聚力和连锁作用来抵抗的。未来的研究将重点放在不同泥浆中摩擦的量化、砖-泥浆相互作用和砖-泥浆组合的 DEM 建模上[27]。

The conducted analysis has demonstrated that the failure of the **refractory mortars** under investigation is caused by a sequence of discrete local events. In shear, these events can take form either of the slip of grains or their tensile de-bonding. On the global shear stress-strain curve, the failure events are marked by peaks of various magnitudes. Under compression, the cracks and pore closure occur due to the rearrangement of grains and their clusters. This produces exponential stress-strain curves where the stiffness increases with load. The mortar porosity can both reduce the global failure stress and promote the crack arrest. The binder influences the grain cohesion by changing the bonding and the friction between the grains. The effects of friction are more pronounced in presence of lateral forces. Materials of higher friction between the grains are stiffer, stronger and have less brittle post failure behaviour. The mortar with immature binder is characterised by intensive crack branching. This, in combination with the effects of porosity, results in high compressibility and low strength and high ability for the crack arrest. In mature binder the crack branching is resisted by the grain cohesion and interlocking. Future research will focus on quantifying the friction in different mortars, the brick-mortar interaction and DEM modelling of the brick-mortar assembly[27].

在**耐火泥浆**（配制批次）中添加高岭土可以改善耐火泥浆的体积密度和黏结强度等性能，但这种改善是有一定添加量限制的。结果表明，高岭土的最佳添加量为40%。增加高岭土的含量，可使耐火泥浆中莫来石、方石、刚玉和玻璃相的形成增加，体积密度、黏结强度、导热系数、比热容和抗热震性提高，孔隙率和吸水率降低。在高温下，随着温度的突然变化，在结合区域内扩展的宏观和微观裂纹的发展和扩大，使得结合强度变弱。可以使用不同的添加剂，如硅酸盐、碳酸盐和长石来改善耐火泥浆的品质[28]。

The addition of kaolin to **refractory mortar** leads to improve the properties of refractory mortar, such as bulk density and bond strength, but this improvement continues to a certain limit of added amount. The best percentage content of added kaolin was found to be 40%. Increasing the percentage content leads to increase the formation of mullite, cristobalite, corundum and glass phases in the refractory mortar (prepared batches) as well as the bulk density, bond strength, thermal conductivity, specific heat capacity and thermal shock resistance and to decrease porosity and water absorption capacity. The bond strength became weak at high work temperatures due to the development and growing of macro-and micro-cracks spreading within the bond regions when sudden changes in temperatures occur. It is possible to use different additives, such as silicates, carbonates and feldspar to improve the quality of refractory mortar[28].

耐火泥浆用于铺设耐火砖和定型耐火制品，其具有以下用途：可以将砖结合成一个坚固的单元使其更能抵抗冲击和应力，可以在稍微不规则的砖表面之间提供缓冲从而使一层砖能牢固地支撑上一层砖，也可以使砖墙不漏气或防止熔渣渗透到接缝中。最好的泥浆是塑料黏土和体积恒定的灰泥的组合，因为单独使用原始的耐火黏土作为泥浆仅限于低温应用。泥浆有时用水稀释，用作耐火墙表面的涂层，以进一步密封接缝或保护墙免受炉中破坏性元素的影响。塑性耐火材料是由压碎的耐火砖或熟料与塑性耐火黏土黏合而成的粗骨料，一般有三种用途。第一种是用于制作成型耐火材料形状，以在坯体状态下在炉中使用；二是形成模压整体式墙体或炉膛结构；第三是修复和修补破损的砖[29]。

Refractory mortars are used in laying refractory bricks and shapes and serve the following purposes: to bond the brick-work into a solid unit so that it will be more resistant to shocks and stresses, to provide a cushion between the slightly irregular surfaces of the bricks so that one course of brick work will have a firm bearing on the course below it, and also to make a wall gastight or to prevent penetration of slag into the joints. The best mortars are combinations of a plastic clay and a volume constant grog, as the use of raw fireclay alone as a mortar is confined to low-temperature application. Mortars are sometimes thinned with water and used as coatings for the face of the refractory walls in order to seal the joints further or to protect the wall from destructive elements in the furnace. Plastic refractory are composed of a coarse aggregate of crushed firebrick or grog bonded with plastic fireclay and are used for three general purposes. The first is for making molded refractory shapes to be used in the furnace in the green state; the second is to form a molded monolithic wall or furnace structure; and the third is to repair and patch worn brick work[29].

高温涂料在许多行业有着广泛的应用。人们在改进涂层技术以保护不同结构方面做了大量的工作。涂层的主要目的是保护部分或全部金属结构。涂料可以分为不同的类别，但有两大类：有机和无机。由于有机涂料的热稳定性较低，无机涂料已被提出用于工业应用。陶瓷涂料是由黏结剂和填料组成的高耐热性无机涂料。无机黏结剂因其热稳定性好而广泛应用于陶瓷涂料的制备[30]。

High-temperature coatings have a wide range of applications in many industries. There has been extensive work on improving the coating technology to protect different structures. The main goal of a coating is to protect structures that are partly or fully metallic. Coatings can be classified into different categories, but there are two major groups: organic and inorganic. Due to the low thermal stability of organic coatings, inorganic coatings have been proposed for industrial applications. Ceramic coatings are inorganic coatings with high thermal resistance and are composed of binders and fillers. In the preparation of ceramic coatings, inorganic binders are used owing to their high thermal stability[30].

纳米陶瓷或涂层的潜在好处包括改进的材料特性，例如由于减小的晶粒尺寸和滑移距离、减小的缺陷尺寸和增强的晶界应力松弛导致的更高的硬度、强度和韧性，即使在环境温度下也是如此。扩散率大大增加，与较大体积的晶界有关。陶瓷或耐火材料的热导率可能会降低，因为来自晶界和其他纳米级特征的声子散射增强。因此，基于蓝晶石和红柱石的**纳米级耐火涂料**在工程工业应用中具有巨大的潜力。一种典型的应用可能是通过提供可以满足防止高温腐蚀要求的纳米颗粒涂层来原位修复高炉[31]。

The potential benefits of nano ceramics or coatings include improved materials characteristics such as higher hardness, strength and toughness resulting from reduced grain size and slip distance, reduced defect size and enhanced grain boundary stress relaxation, even at ambient temperature. Diffusivity is greatly increased, associated with a larger volume of grain boundaries. Thermal conductivity in ceramics or refractories may be reduced because of enhanced phonon scattering from grain boundaries and other nanoscale features. Thus **nano sized refractory coatings** based on kyanite and andulsite have great potential for engineering industrial applications. One typical application may be in-situ repairs of blast furnace by providing coatings of nanoparticles that can satisfy requirements of protection from high temperature corrosion[31].

这涉及固体原材料的雾化或汽化，并将材料沉积到基材上以形成涂层。尽管PVD方法通常需要更昂贵的设备，更多的零件准备、系统维护和更长的加工时间，但PVD是一种可行的**耐火硬质涂层**工艺。在某些情况下，PVD是唯一能够沉积所需材料涂层的方法。当考虑到所有因素时，PVD的总成本通常可以低于其他涂层方法。在选择PVD工艺之前，必须评估的主要考虑因素包括期望的最终结果、基材性能、清洁和制备、固定、温度和涂层速率。钼、钽、钨、钌等难熔金属及其合金非常适合用PVD沉积[32]。

This involves the atomization or vaporization of material from a solid source and deposition of the material onto the substrate to form a coating. Although PVD methods typically require more expensive equipment, more parts preparation, more system maintenance and greater processing times than many alternative processes, PVD is a viable coating process for **refractory and hard coatings**. In some instances, PVD is the only method capable of depositing a coating of the desired material. The overall cost of the PVD often can be less than other coating methods when all the factors are taken into account. The primary considerations that must be evaluated before selecting a PVD process include the desired end result, substrate properties, cleaning and preparation, fixturing, temperature and coating rates. Refractory metals such as Mo, Ta, W, Ra and their alloys are well suited for deposition by PVD[32].

目前，消失模铸造工艺的一个严重缺陷是铸件中容易形成气孔。由于聚苯乙烯图案在砂型中燃烧而产生空洞或气孔。气体在铸造过程中析出。为了克服吹孔问题，将 EPC 工艺与真空密封成型工艺相结合。真空技术应用于消失模铸造过程中，吸收了分解气体，提高了铸件质量。开发的混合工艺被称为真空辅助蒸发模铸造（VAEPC）工艺。研究人员对利用无黏结砂型在引入熔融金属过程中包含图案产生了兴趣。使用特殊黏合剂将组装的图案段黏合成完整的图案。如果要生产可接受的无缺陷铸件，则要求这些黏合剂具有某些特性。然后将单个图案的集群组装并连接到使用类似黏合剂的中央垂直浇注系统。然后将**耐火涂层**涂在图案簇的整个表面，并在低温烘箱中干燥[33]。

At the present time, one serious limitation of EPC process is the formation of blow holes in the castings. The blow holes or gas porosities are generated due to the burning of polystyrene pattern in the sand mold. The gases evolve in the casting. To overcome the problem of blow holes, EPC process is combined with the vacuum sealed moulding process. The vacuum applied to EPC mould draws the decomposed gases and improves the casting quality. The developed hybrid process has been termed as vacuum assisted evaporative pattern casting (VAEPC) process. Researchers have developed interest in utilizing unbonded sand mold to contain the pattern during introduction of molten metal. Special adhesives are utilized to bond the assembled pattern segments into a completed pattern. These adhesives are required to exhibit certain characteristics if acceptable defect free castings are to be produced. Clusters of individual patterns are then assembled and joined to a central vertical gating system utilizing similar adhesives. **Refractory coating** is then applied to the total surface of the pattern cluster and dried in a low temperature oven[33].

参考文献

[1] Li Ye, Zhu Lingling, Liu Kun, et al. Effect of curing temperature on volume stability of CAC-bonded alumina-based castables [J]. Ceramics International, 2019, 45 (9): 12066-12071.

[2] Zhu Boquan, Song Yanan, Li Xiangcheng, et al. Synthesis and hydration kinetics of calcium aluminate cement with micro $MgAl_2O_4$ spinels [J]. Materials Chemistry and Physics, 2015, 154: 158-163.

[3] Zhang Yang, Ye Guotian, Gu Wenjing, et al. Conversion of calcium aluminate cement hydrates at 60 ℃ with and without water [J]. Journal of the American Ceramic Society, 2018, 101 (7): 2712-2717.

[4] Li Ye, Guo Liu, Ding Dafei, et al. Effect of grinding on the hydration of hydratable alumina and properties of hydratable alumina-bonded castables [J]. Ceramics International, 2021, 47 (5): 6505-6512.

[5] Yang Yufei, Liu Hao, Wang Zhoufu, et al. Microstructure and enhanced slag resistance of Al_2O_3-SiC-C refractory castables with addition of ammonium metatungstate [J]. Ceramics International, 2023, 49 (14): 23558-23566.

[6] Guo Liu, Song Shengqiang, Mu Yuandong, et al. Inhibited Cr(Ⅵ) formation in Cr_2O_3-containing refractory castable using reactive MgO as hydraulic binder [J]. Journal of the European Ceramic Society, 2022, 42 (16): 7656-7666.

[7] Xu Lei, Liu Yang, Chen Min, et al. An accurate correlation between high-temperature performance and cement content of the high-alumina refractory castables [J]. Ceramics International, 2022, 48 (15): 22601-22607.

[8] Wang Yulong, Li Xiangcheng, Chen Pingan, et al. Matrix microstructure optimization of alumina-spinel castables and its effect on high temperature properties [J]. Ceramics International, 2018, 44 (1): 857-868.

[9] Zhang Wei, Meng Qian, Dai Wenyong. Research on application of kyanite in plastic refractory [J]. Chinese Journal of Geochemistry, 2013, 32 (3): 326-330.

[10] Brulin J, Rekik A, Josserand L, et al. Characterization and modelling of a carbon ramming mix used in high-temperature industry [J]. International Journal of Solids and Structures, 2011, 48 (5): 854-864.

[11] Tang Hu, Jia Zhenggang, Li Bing, et al. Enhanced properties of tailored alumina-magnesia-based dry ramming mixes by calcium magnesium aluminate (CMA) [J]. Materials, 2023, 16 (4): 1707.

[12] Luo, Xudong, Zhang Guodong, Xie Zhipeng, et al. Effects of bauxite on property of Al_2O_3-SiO_2 plastic refractory [J]. Non-metallic Mines, 2015, 38 (1): 29-31.

[13] Wei Huixian, Jiang Yanwei, Zhang Qiang, et al. Effect of boric acid on properties of high-alumina phosphate-bonded plastic refractory materials [J]. Journal of the Chinese Chemical Society, 2018, 65 (9): 1053-1059.

[14] Pan Dunxiang, Zhao Huizhong, Zhang Han, et al. Corrosion mechanism of spray refractory in COREX slag with varying basicity [J]. Ceramics International, 2019, 45 (18): 24398-24404.

[15] Kovarik O, Materna A, Siegl J, et al. Fatigue crack growth in plasma-sprayed refractory materials [J]. Journal of Thermal Spray Technology, 2019, 28: 87-97.

[16] Cao Feng, Meng Qingmin, Long Shigang, et al. Effect of rheological behavior of particle-water suspensions on properties of gunned refractory for blast furnace [J]. Journal of Iron and Steel Research, International, 2006, 13 (1): 10-13.

[17] Zhang Wei, Yao Yinzhi, Dai Wenyong. Effects of kyanite additions on properties of bauxite-based gunning refractory [J]. Non-Metallic Mines, 2009, 32 (3): 27-30.

[18] 顾华志, 张文杰. 不定形耐火材料节能化研究进展 [J]. 耐火材料, 2012, 46 (3): 161-168.

[19] Cai Manfei, Liang Yonghe, Yin Yucheng, et al. Effect of citric acid on the hydration process of colloidal silica-bonded magnesia gunning materials [J]. Ceramics International, 2019, 45 (12): 15514-15519.

[20] Steindl F R, Mittermayr F, Sakoparnig M, et al. On the porosity of low-clinker shotcrete and accelerated pastes [J]. Construction and Building Materials, 2023, 368: 130461.

[21] Xu Lei, Gao Song, Chen Min, et al. Improvement in microstructure and thermo-mechanical properties of MgO-based dry vibratable material by addition of Fe [J]. Materials Chemistry and Physics, 2020, 253: 123368.

[22] 徐振东, 徐素英, 刘永杰. 中间包工作层用干式捣打料的研制 [J]. 山东冶金, 2005, 27 (5): 43-45.

[23] 商思凯, 孙义, 王大博. 中间包长寿命工作层新技术的开发与应用 [J]. 黑龙江冶金, 2005 (4): 25-26, 28.

[24] 饶康. 基质调控对感应炉用铝镁质干式料性能的影响 [D]. 武汉: 武汉科技大学, 2020.

[25] Guo Mengzhen, Liang Yonghe, Cai Manfei, et al. Improved slag resistance of taphole clay due to in situ formation of TiCN from ferrotitanium slag [J]. Ceramics International, 47 (16): 23630-23636.

[26] Li Xueyin, Fang Minghao, Liu Yangai, et al. Fe-Sialon-Ti(C,N) composites from carbothermal reduction-nitridation of low-priced minerals and their application in taphole clay refractories [J]. Ceramics International, 2014, 40 (7): 9709-9714.

[27] Andreev K, Verstrynge E, Wevers M. Effect of binding system on the compressive behaviour of refractory mortars [J]. Journal of the European Ceramic Society, 2014, 34 (13): 3217-3227.

[28] Mahdi D F. Preparation of refractory mortar from Iraqi raw materials [J]. Iraqi Journal of Applied Physics, 2015, 11 (2): 79-82.

[29] Kadhum M, Jaffer H A. Study of the thermal durability of refractory mortar prepared from local clay mixed with different percentage of silica [J]. Iraqi Journal of Science, 2013, 54 (4): 1096-1101.

[30] Khosravanihaghighi A, Pakshi M. Effects of SiC particle size on electrochemical and mechanical behavior of SiC-based refractory coatings [J]. Journal of the Australian Ceramic Society, 2017, 53: 909-915.

[31] Sanjai S G, Pinto R, Ramaswamy P. Plasma sprayed nano refractory coatings [C] //IOP Conference Series: Materials Science and Engineering. IOP Publishing, 2019, 577 (1): 012100.

[32] Dahotre N B, Kadolkar P, Shah S. Refractory ceramic coatings: processes, systems and wettability/adhesion [J]. Surface and Interface Analysis, 2001, 31 (7): 659-672.

[33] Kumar S, Kumar P, Shan H S. Characterization of the refractory coating material used in vacuum assisted evaporative pattern casting process [J]. Journal of Materials Processing Technology, 2009, 209 (5): 2699-2706.

7 耐火材料制品常用词汇术语查询

7.1 汉英对照

A

Al_2O_3-SiC-C 砖 alumina-SiC-C brick

Al_2O_3-SiO_2-SiC-C 质炮泥 Al_2O_3-SiO_2-SiC-C mud

B

白云石耐火材料 dolomite refractory; dolomitic refractory

白云石耐火砖 dolomite refractory brick

板状刚玉 tabular corundum

半 SiC 质制品 semi-silicon carbide products

半硅质耐火材料 semi-silica refractories

半硅质原料 semi-silica raw materials

半石墨碳砖 semi graphite carbon brick

半稳定白云石耐火砖 semi-stablized dolomite refractory brick

半再结合镁铬砖 semi-rebonded magnesia chrome brick

焙烧碳砖 roasted carbon brick

玻璃棉 glass wool

不烧高铝砖 unfired high alumina bricks

不烧隔热耐火材料 unroasted thermal insulating refractory materials

不烧轻质黏土砖 unfired lightweight clay bricks

C

超低水泥耐火浇注料 ultra-low cement refractory castable

超细二氧化硅微粉 ultrafine silica powders

长水口 long nozzles

盛钢桶用高铝砖 high alumina bricks for casting ladle

D

单晶碳化硅 single crystal silicon carbide

氮化硅基纳米复合材料 silicon nitride-based nanocomposites

氮化硅结合碳化硅制品 silicon nitride bonded silicon carbide products

氮化硅陶瓷 silicon nitride ceramics; Si_3N_4 ceramics

氮化铝陶瓷 aluminium nitride ceramic

氮化物结合 SiC 制品 nitride bonded SiC

捣打料 ramming refractory

低气孔耐火黏土砖 low porosity fireclay bricks

低蠕变高铝砖 low creep high alumina bricks

低碳镁碳砖 low carbon magnesia carbon brick

低碳耐火材料 low carbon refractory

电炉底用镁钙铁质干式振捣料 alkaline dry vibratable refractory for electric furnace bottom

电炉碳砖 electric furnace carbon brick
电熔镁砂（电熔氧化镁）fused magnesia
定径水口 metering nozzles
多晶碳化硅 polycrystalline silicon carbide
多晶氧化铝基纤维 polycrystalline alumina-based fibers
多晶氧化铝纤维 polycrystalline alumina fiber
多孔氮化硼 porous boron nitride
多孔隔热耐火材料 porous refractory thermal insulating materials
多孔硅酸钙 porous calcium silicate
多孔轻质黏土砖 porous lightweight clay bricks
多孔石墨 porous graphite
多孔碳化硅陶瓷 porous silicon carbide ceramics
多孔氧化铝陶瓷 porous alumina ceramic
多熟料耐火黏土砖 high grog fireclay bricks

E

二硅化钼 molybdenum disilicide；$MoSi_2$
二硅化钼金属间化合物 molybdenum disilicide intermetalic
二铝酸钙 calcuim dialuminate（$CaO \cdot 2Al_2O_3$，CA_2）
二氧化硅结合碳化硅制品 silicon dioxide bonded silicon carbide products
二元硼化物陶瓷 binary boride cermets

F

矾土基高铝砖 bauxite-based high alumina bricks
矾土基莫来石 bauxite-based mullite
矾土基莫来石均质料 bauxite-based homogenized mullite grogs
矾土耐火材料 bauxite refractories
反应结合氮化硅 reaction-bonded silicon nitride
反应烧结氮化硼 reactive sintered boron nitride
反应烧结多孔氮化硅陶瓷 reaction-bonded porous silicon nitride ceramics
反应烧结锆莫来石复合材料 reaction-sintered zirconia-mullite composites
方镁石-镁橄榄石轻质隔热耐火材料 periclase-forsterite lightweight heat-insulating refractories
方镁石-镁铝尖晶石耐火砖 periclase-magnesium aluminate spinel refractory（PMAS）
方镁石-镁铝尖晶石砖 periclase-magnesia alumina spinel bricks
非晶碳化硅 amorphous silicon carbide
粉煤灰 fly ash
粉煤灰漂珠 fly ash floating beads
粉煤灰悬浮液 fly ash suspension
氟化石墨 graphite fluoride
富铝铝酸钙耐火材料 alumina-rich calcium aluminate refractory

G

钙长石基隔热耐火砖 anorthite based insulating firebricks
钙长石轻质隔热砖 lightweight anorthite insulation bricks
钙质粉煤灰 calcareous fly ash
感应炉用碱性干式振捣料 alkaline dry vibratable refractory for induction fuenace
干式耐火振捣料 dry vibratable refractory
刚玉尖晶石透气砖 corundum-spinel purging plug
刚玉尖晶石质耐火材料 corundum-spinel refractories
刚玉镁铝尖晶石 corundum-magnesia alumina spinel
刚玉莫来石砖 corundum-mullite bricks
刚玉耐火材料 corundum refractories
刚玉耐磨可塑料 corundum wearable plastic refractory
刚玉质干式振捣料 corundum dry vibratable refractory
刚玉质透气砖 corundum purging plug
刚玉砖 corundum bricks
高发射率涂层 high emissivity coatings
高隔热耐火材料 high-refractory thermal insulating materials
高炉用复合棕刚玉砖 composite brown corundum

bricks for blast furnace
高炉用碳砖 carbon brick for blast furnace
高铝耐火砖 high-alumina refractory bricks
高铝塞头砖 high-alumina stopper
高铝-碳化硅-碳质捣打料 Al_2O_3-SiC-C（ASC）ramming refractory
高铝-碳化硅-碳质喷射耐火材料 Al_2O_3-SiC-C（ASC）gunning refractory
高铝质耐火材料 high-alumina refractory
高铝质制品 high alumina products
高铝砖 high alumina bricks
高密度硅砖 high density silica bricks
高温红外辐射节能涂层 high temperature infrared radiation energy saving coating
高温模压碳砖 high-temperature mould pressing carbon brick
高氧化锆熔铸耐火材料 high zirconia fused-cast refractories
高氧化锆砖 high zirconia brick
高氧化铬耐火材料 high chrome oxide refractories
锆刚玉坩埚 corundum-zircon crucible
锆刚玉耐火材料 corundum-zircon refractories
锆刚玉砖 zircon corundum brick
锆英石质耐火捣打料 zircon ramming refractory
隔热耐火材料 refractory thermal insulation materials；insulation refractories
铬刚玉砖 chrome-corundum bricks
硅化钨薄膜 tungsten silicide films
硅莫砖 guimo bricks
硅酸钙基材料 calcium silicate-based materials
硅酸铝耐火纤维 aluminium silicate refractory fibre
硅酸铝质干式振捣料 aluminium silicate dry vibratable refractory
硅酸铝质隔热耐火泥浆 alumina-silica insulating refractory mortars
硅酸铝质喷射耐火材料 aluminium silicate gunning refractory
硅线石砖 sillimanite bricks
硅藻土 diatomite
硅藻土复合水凝胶 the diatomite composite hydrogel
硅质粉煤灰 siliceous fly ash
硅质干式振捣料 silica dry vibratable refractory
硅砖 silica brick

H

含碳耐火材料 carbon-bearing refractory；carbon containing refractory
滑动水口 sliding nozzle；sliding gate
化学结合镁铬砖 chemical-bonded magnesia chrome brick；unfired magnesia chrome brick
化学结合镁砖 chemical bonded magnesite brick
环保轻质黏土砖 environmentally friendly lightweight clay bricks
火焰喷补耐火材料 flame gunning refractory

J

尖晶石-镁橄榄石 spinel-forsterite
尖晶石镁锆砖 magnesia-spinel-zirconia brick
碱式喷射耐火材料 alkaline gunning refractory
浇钢用隔热板 insulation boards for steel casting
焦炉硅砖 coke oven silica bricks
焦炉用半硅砖 semi-silica brick for coke oven
焦油结合镁砖 tar-bonded magnesite bricks
节能涂料 energy-saving coatings
浸入式水口 submerged entry nozzles

K

抗剥落高铝砖 spalling resistant high-alumina bricks
抗蠕变高铝砖 creep-resistant high-alumina bricks
可水合氧化铝结合浇注料 hydratable alumina bonded castable

L

立方碳化硅 cubic silicon carbide
磷酸浸渍耐火黏土砖 phosphoric acid immersed fireclay bricks

磷酸盐结合浇注料 phosphate-bonded castable
磷酸盐结合可塑料 phosphate-bonded castable
鳞片石墨 flake graphite
六铝酸钙 calcium hexaluminate（$CaO \cdot 6Al_2O_3$，CA_6）
铝锆碳质耐火材料 alumina zirconia carbon refractory
铝锆碳砖 Al_2O_3-ZrO_2-C brick
铝镁碳质耐火材料 alumina magnesia carbon refractory
铝-镁质捣打料 alumina-magnesia ramming refractory
铝-镁质浇注料 alumina-magnesia castable
铝酸钙耐火材料 calcium aluminate refractories
铝酸钙水泥 calcium aluminate cement（CAC）
铝酸钙水泥结合浇注料 calcium aluminate cement-bonded castable
铝碳质耐火材料 alumina carbon refractory
铝碳砖 alumina carbon brick；Al_2O_3-C brick

M

镁钙碳砖 magnesia calcia carbon brick
镁橄榄石耐火砖 forsterite refractory brick
镁橄榄石砖 magnesia forsterite bricks
镁锆耐火材料 magnesia-zirconia refractory
镁锆砖 magnesia zirconia brick；magnesia-zirconia brick
镁铬耐火材料 magnesia-chrome refractory
镁铬砖 magnesia chrome brick；magnesia-chromite brick
镁硅砖 magnesia-silica brick；high-silica magnesite brick
镁基耐火材料 magnesia-based refractory
镁铝铬复合尖晶石砖 magnesia alumina chrome composite spinel bricks
镁铝尖晶石砖 magnesia-spinel bricks
镁铝钛砖 magnesia-alumina-titania bricks
镁铝碳砖 magnesia-alumina-carbon（MAC）bricks
镁碳砖 magnesia carbon brick；MgO-C brick
镁铁铝尖晶石砖 magnesia hercynite bricks
镁质白云石耐火材料 magnesite-dolomite refractory；magnesia-dolomite refractory
镁质干式振捣料 magnesia dry vibratable refractory
镁砖 magnesia brick
免烧高铝砖 unfired high alumina bricks
莫来石-SiC-O-Sialon 复合材料 mullite-SiC-O-Sialon composites
莫来石刚玉格子砖 mullite-corundum checker bricks
莫来石刚玉制品 mullite-corundum products
莫来石刚玉质耐火材料 mullite-corundum refractory
莫来石结合碳化硅制品 mullite bonded silicon carbide products
莫来石-堇青石棚板 mullite-cordierite decks
莫来石-堇青石窑具 mullite-cordierite kiln furniture
莫来石-堇青石质匣钵 mullite-cordierite saggar
莫来石熟料 mullite chamotte
莫来石陶瓷 mullite ceramics
莫来石系轻质砖 lightweight mullite bricks
莫来石-氧化锆质耐火材料 mullite-zirconia refractories
莫来石铸块 mullite blocks
莫来石砖 mullite bricks

N

纳米镁橄榄石/纳米镁铝尖晶石粉 nano forsterite/nano magnesium aluminate spinel powders
耐火挤压料 refractory mud
耐火浇注料 castable refractory
耐火可塑料 plastic refractory
耐火泥浆 refractory mortar
耐火黏土 fire clay
耐火黏土绝热板 fireclay insulating board
耐火黏土砖 fireclay bricks
耐火涂料（涂抹料）refractory coating
耐火压注料 injection refractory
耐火氧化物空心球 hollow spheres of refractory
耐碱耐火浇注料 alkaline resistant refractory castable
耐碱耐火涂料 alkaline resistant refractory coating

耐磨耐火浇注料 wearing resistant refractory castable
耐酸耐火浇注料 acid resistant refractory castable
耐酸耐火涂料 acid resistant refractory coating
黏土格子砖 fireclay checker bricks
黏土结合可塑料 clay bonded plastic refractory
黏土结合碳化硅制品 clay bonded silicon carbide products
黏土袖砖 fireclay sleeve
黏土砖 clay bricks

P

泡沫碳化硅 foam silicon carbide
喷补用耐火材料 gunned refractory
喷射耐火材料 gunning refractory
硼化锆-铬镍合金 zirconium boride-nichrome
硼化锆颗粒 zirconium boride particles
硼化锆-硼化物 silicon-boron-zirconium boride
硼化锆涂层 zirconium boride coating
硼化物基金属陶瓷 boride-based cermets
膨胀石墨 expanded graphite; exfoliated graphite; EG
膨胀珍珠岩 expanded perlite
膨胀蛭石 expanded vermiculite
漂珠 floating bead
漂珠陶瓷 floating bead ceramics
普通耐火黏土砖 common fireclay refractory bricks

Q

七铝酸十二钙 dodecacalcium hepta-aluminate ($12CaO \cdot 7Al_2O_3$, $C_{12}A_7$)
气凝胶超级隔热材料 aerogel super insulation materials
气硬性耐火泥浆 air setting jointing material ; refractory mortar of air setting
轻烧镁砂（轻烧氧化镁）caustic magnesite (light-burned magnesia)
轻质（隔热）耐火浇注料 lightweight refractory castable
轻质 Al_2O_3-MgO-C 耐火材料 lightweight Al_2O_3-MgO-C refractories
轻质高铝砖 lightweight high alumina bricks
轻质隔热硅砖 lightweight insulation silicon bricks
轻质隔热耐火材料 lightweight insulation refractories
轻质隔热黏土砖 lightweight insulating fireclay bricks
轻质硅砖 lightweight silica bricks
轻质镁橄榄石砖 lightweight forsterite brick
轻质莫来石基莫来石-碳化硅砖 lightweight mullite-based mullite-SiC bricks
轻质黏土砖 lightweight clay bricks
轻质微孔镁基耐火材料 lightweight microporous magnesia-based refractories
轻质氧化镁 lightweight magnesia
轻质砖 lightweight bricks

R

热风炉硅砖 hot-blast stove silica bricks
热辐射涂料 heat radiative coating
热硬性耐火泥浆 refractory mortar of heat setting
熔融镁铬砖 fused magnesia chrome brick; fused grain magnesia chrome brick
熔融莫来石 fused mullite
熔融莫来石粉末 fused mullite powder
熔融莫来石颗粒 fused mullite particles
熔融氧化锆-氧化铝-二氧化硅耐火材料 fused zirconia-alumina-silica material (AZS) refractory
熔铸刚玉制品 fused cast corundum refractory products
熔铸刚玉砖 fused cast corundum bricks
熔铸锆刚玉砖 fused cast zirconia-corundum bricks
熔铸铬刚玉砖 fused cast chrome-corundum bricks
熔铸辉石刚玉耐火材料 fused-cast baddeleyite-corundum refractories
熔铸铝硅锆（AZS）耐火材料 fused-cast AZS refractories
熔铸氧化锆耐火材料 fused cast zirconia refractory

S

塞隆结合碳化硅制品 Sialon bonded silicon carbide products

三元硼化物基陶瓷 ternary boride-based cermet

烧成镁砖 fired magnesia brick

烧结氮化硅 sintered silicon nitride

烧结锆莫来石耐火制品 sintered zirconia mullite refractory products

烧结锆莫来石砖 zirconium mullite

烧结镁砂（死烧镁砂）sintered（dead burned）magnesia

烧结莫来石制品 sintered mullite products

烧结氧化锆耐火制品 sintered zirconia refractory products

烧结氧化锆陶瓷 sintered zirconia ceramics

渗硅反应 SiC 质制品 silicon infiltration reaction silicon carbide products

石灰结合硅质耐火材料 lime-bonded silica refractory

石蜡/膨胀珍珠岩材料 paraffin/expanded perlite material

石棉 asbestos

石棉布复合纳米材料 asbestos cloth nanocomposite

石棉酚醛复合材料 asbestos-phenolic composite

石墨（质）耐火材料 graphite refractory

石墨（质）制品 graphite articles

石墨板 graphite sheet

石墨棒 graphite bar

石墨大理石 graphite marble

石墨电极 graphite electrode

石墨发热体 graphite heater

石墨粉 graphite powder

石墨风口砖 graphite tuyere block

石墨坩埚 graphite dust；graphite powder；powdered graphite；graphite crucible

石墨管 graphite tube

石墨环 graphite annulus

石墨火泥 graphite mortar

石墨模型 graphite mould

石墨黏土砖 carbon-clay brick

石墨填料 graphite packing

石墨纤维 graphite fibers

石墨相氮化碳 graphitic carbon nitride；C_3N_4

石墨压模 graphite compression mould

石墨阳极 graphite anode；graphite cathode

石墨质熔池 graphite bath

石墨舟 graphite boat

水泥窑用硅莫红砖 GMH bricks for cement kilns

水硬性耐火泥浆 refractory mortar of hydraulic setting

酸化石墨 acidified graphite

T

碳棒 carbon bar

碳复合耐火材料 carbon composite refractory

碳化硅晶须 silicon carbide whiskers

碳化硅膜 silicon carbide film

碳化硅纳米管 silicon carbide nanotubes

碳化硅纳米片 silicon carbide nanosheets

碳化硅纳米线 silicon carbide nanowires

碳化硅陶瓷基复合材料 silicon carbide ceramic matrix composite material

碳化硅纤维 silicon carbide fiber

碳化物耐火材料 carbide refractory

碳素捣打料 carbon ramming mix

碳纤维 carbon fiber；carbon filament

碳质火泥 carbon mortar

碳质内衬 carbon lining

碳质泥料 carbon loam；carbon mass；carbon mix

碳质黏土 carbonaceous clay

碳砖（碳素砖）carbon brick

碳砖糊料 carbon brick paste

陶瓷气凝胶 ceramic aerogels

特种耐火泥浆 special refractory mortar

天然硅藻土 natural diatomite

铁硅化物纳米颗粒 iron silicide nanoparticles

透明氧化铝陶瓷 transparent alumina ceramics

透气元件 porous plugs

$TiZrO_4$ 空心球 $TiZrO_4$ hollow spheres

W

微波烧结氧化锆 microwave sintered zirconia

微晶石墨 microcrystalline graphite

微孔刚玉砖 micropore corundum bricks

微孔硅酸钙 microporous calcium silicate

微孔耐火材料 microporous refractories

微孔轻质隔热材料 microporous lightweight insulator materials

微孔碳砖 micropore carbon brick

温石棉 chrysotile

稳定性白云石耐火砖 stabilized dolomite refractory brick

无水泥耐火浇注料 cement free refractory castable

无水炮泥 waterless mud

X

纤维多孔氧化锆陶瓷 fibrous porous zirconia ceramics

旋涡浸入式水口 the submerged entry nozzles with swirling flow

Y

压注料 injection mix

亚晶锆石纳米气凝胶 hypocrystalline zircon nanofibrous aerogels

岩棉 rock wool

氧氮化硅和复相氮化物结合碳化硅制品 silicon oxy nitride and compound phase nitride bonded silicon carbide products

氧化钙坩埚 calcia crucible; calcium oxide crucible; CaO crucible

氧化钙制品 calcium oxide products; calcia products; CaO products

氧化锆耐火材料 zirconia refractory

氧化锆纤维 zirconia fiber

氧化锆制品 zirconia products

氧化铬制品 chrome oxide products

氧化铝坩埚 alumina crucible

氧化铝基耐火陶瓷 alumina-based refractory ceramics

氧化铝空心球 alumina hollow spheres

氧化铝耐火砖 alumina firebricks; alumina refractory bricks

氧化铝轻质保温砖 alumina lightweight insulation brick

氧化铝-碳化硅-碳质浇注料 Al_2O_3-SiC-C (ASC) castable

氧化铝制品 alumina products

氧化镁坩埚 magnesia crucible

氧化镁-镁铝尖晶石耐火材料 magnesia-magnesium aluminate spinel refractory

氧化镁耐火材料制品 magnesia refractory products

氧化镁氧化锆复合材料 magnesia zirconia composite

氧化镁-氧化锆共混物 magnesia-zirconia co-clinker

氧化镁-氧化锆陶瓷 MgO-ZrO_2 ceramic

氧化镁制品 magnesia products

氧化物结合 SiC 制品 oxide bonded SiC

冶金镁砂 metallurgical magnesia

一铝酸钙 calcuim monoaluminate (CaO · Al_2O_3)

优质耐火黏土砖 high-duty fireclay bricks

有水炮泥 taphole mud

预焙碳块 prebaked carbon block

预合成氧化镁-氧化锆 presynthesized magnesia-zirconia

远红外涂料 far-infrared coating

Z

再结合电熔刚玉砖 rebonded electrically fused corundum bricks

再结合镁锆砖 re-bonded magnesia zirconia brick

再结合镁铬砖 re-bonded magnesia chrome brick

再结合镁砖 rebonded magnesite brick

再结合烧结刚玉砖 rebonded sintered corundum bricks

再生电熔粒状镁铬耐火砖 reconstituted fused-grain magnesia-chrome refractory brick
整体塞棒 monolithic stopper
直接结合镁铬砖 direct-bonded magnesia chrome brick
致密黏土砖 dense fireclay bricks
致密烧结刚玉陶瓷 dense sintered corundum ceramics
中间包用碱性干式振捣料 alkaline dry vibratable refractory for tundish
重结晶 SiC 制品 resystallized silicon carbide products
铸造 AZS 耐火材料 casting AZS refractories
自焙碳砖 self-baking carbon brick
自结合 SiC 制品 self-bonded silicon carbide products
*
Ⅲ等高铝砖 grade Ⅲ high alumina bricks
Ⅱ等高铝砖 grade Ⅱ high alumina bricks
Ⅰ等高铝砖 grade Ⅰ high alumina bricks
β-Sialon 结合刚玉碳化硅复合材料 β-Sialon bonded corundum-SiC composites
β-Sialon 结合刚玉砖 β-Sialon bonded corundum bricks
β-SiC 结合 SiC 制品 β-SiC bonded silicon carbide products

7.2 英汉对照

A

acid resistant refractory castable 耐酸耐火浇注料
acid resistant refractory coating 耐酸耐火涂料
acidified graphite 酸化石墨
aerogel super insulation materials 气凝胶超级隔热材料
air setting jointing material ; refractory mortar of air setting 气硬性耐火泥浆
Al_2O_3-SiC-C (ASC) castable 氧化铝-碳化硅-碳质浇注料
Al_2O_3-SiC-C (ASC) gunning refractory 高铝-碳化硅-碳质喷射耐火材料
Al_2O_3-SiC-C (ASC) ramming refractory 高铝-碳化硅-碳质捣打料
Al_2O_3-SiO_2-SiC-C mud Al_2O_3-SiO_2-SiC-C 质炮泥
Al_2O_3-ZrO_2-C brick 铝锆碳砖
alkaline dry vibratable refractory for electric furnace bottom 电炉底用镁钙铁质干式振捣料
alkaline dry vibratable refractory for induction fuenace 感应炉用碱性干式振捣料
alkaline dry vibratable refractory for tundish 中间包用碱性干式振捣料
alkaline gunning refractory 碱式喷射耐火材料
alkaline resistant refractory castable 耐碱耐火浇注料
alkaline resistant refractory coating 耐碱耐火涂料
alumina carbon brick; Al_2O_3-C brick 铝碳砖
alumina carbon refractory 铝碳质耐火材料
alumina crucible 氧化铝坩埚
alumina firebricks 氧化铝耐火砖
alumina hollow spheres 氧化铝空心球
alumina lightweight insulation brick 氧化铝轻质保温砖
alumina magnesia carbon refractory 铝镁碳质耐火材料
alumina products 氧化铝制品
alumina refractory bricks 氧化铝耐火砖
alumina zirconia carbon refractory 铝锆碳质耐火材料
alumina-based refractory ceramics 氧化铝基耐火

陶瓷

alumina-magnesia castable 铝-镁质浇注料

alumina-magnesia ramming refractory 铝-镁质捣打料

alumina-rich calcium aluminate refractory 富铝铝酸钙耐火材料

alumina-SiC-C brick Al_2O_3-SiC-C 砖

alumina-silica insulating refractory mortars 硅酸铝质隔热耐火泥浆

aluminium nitride ceramic 氮化铝陶瓷

aluminium silicate dry vibratable refractory 硅酸铝质干式振捣料

aluminium silicate gunning refractory 硅酸铝质喷射耐火材料

aluminium silicate refractory fibre 硅酸铝耐火纤维

amorphous silicon carbide 非晶碳化硅

anorthite based insulating firebricks 钙长石基隔热耐火砖

asbestos 石棉

asbestos cloth nanocomposite 石棉布复合纳米材料

asbestos-phenolic composite 石棉酚醛复合材料

B

bauxite refractories 矾土耐火材料

bauxite-based high alumina bricks 矾土基高铝砖

bauxite-based homogenized mullite grogs 矾土基莫来石均质料

bauxite-based mullite 矾土基莫来石

binary boride cermets 二元硼化物陶瓷

boride-based cermets 硼化物基金属陶瓷

C

calcareous fly ash 钙质粉煤灰

calcia crucible; calcium oxide crucible; CaO crucible 氧化钙坩埚

calcium aluminate cement（CAC）铝酸钙水泥

calcium aluminate cement-bonded castable 铝酸钙水泥结合浇注料

calcium aluminate refractories 铝酸钙耐火材料

calcium hexaluminate（$CaO \cdot 6Al_2O_3$，CA_6）六铝酸钙

calcium oxide products; calcia products; CaO products 氧化钙制品

calcium silicate-based materials 硅酸钙基材料

calcuim dialuminate（$CaO \cdot 2Al_2O_3$，CA_2）二铝酸钙

calcuim monoaluminate（$CaO \cdot Al_2O_3$）一铝酸钙

carbide refractory 碳化物耐火材料

carbon bar 碳棒

carbon brick 碳砖（碳素砖）

carbon brick for blast furnace 高炉用碳砖

carbon brick paste 碳砖糊料

carbon composite refractory 碳复合耐火材料

carbon fiber; carbon filament 碳纤维

carbon lining 碳质内衬

carbon loam; carbon mass; carbon mix 碳质泥料

carbon mortar 碳质火泥

carbon ramming mix 碳素捣打料

carbonaceous clay 碳质黏土

carbon-bearing refractory; carbon containing refractory 含碳耐火材料

carbon-clay brick 石墨黏土砖

castable refractory 耐火浇注料

casting AZS refractories 铸造 AZS 耐火材料

caustic magnesite（light-burned magnesia）轻烧镁砂（轻烧氧化镁）

cement free refractory castable 无水泥耐火浇注料

ceramic aerogels 陶瓷气凝胶

chemical bonded magnesite brick 化学结合镁砖

chemical-bonded magnesia chrome brick; unfired magnesia chrome brick 化学结合镁铬砖

chrome oxide products 氧化铬制品

chrome-corundum bricks 铬刚玉砖

chrysotile 温石棉

clay bonded plastic refractory 黏土结合可塑料
clay bonded silicon carbide products 黏土结合碳化硅制品
clay bricks 黏土砖
coke oven silica bricks 焦炉硅砖
common fireclay refractory bricks 普通耐火黏土砖
composite brown corundum bricks for blast furnace 高炉用复合棕刚玉砖
corundum bricks 刚玉砖
corundum dry vibratable refractory 刚玉质干式振捣料
corundum purging plug 刚玉质透气砖
corundum refractories 刚玉耐火材料
corundum-spinel refractories 刚玉尖晶石质耐火材料
corundum wearable plastic refractory 刚玉耐磨可塑料
corundum-magnesia alumina spinel 刚玉镁铝尖晶石
corundum-spinel purging plug 刚玉尖晶石透气砖
corundum-zircon crucible 锆刚玉坩埚
corundum-zircon refractories 锆刚玉耐火材料
corundun-mullite bricks 刚玉莫来石砖
creep-resistant high-alumina bricks 抗蠕变高铝砖
cubic silicon carbide 立方碳化硅

D

dense fireclay bricks 致密黏土砖
dense sintered corundum ceramics 致密烧结刚玉陶瓷
diatomite 硅藻土
direct-bonded magnesia chrome brick 直接结合镁铬砖
dodecacalcium hepta-aluminate（$12CaO \cdot 7Al_2O_3$，$C_{12}A_7$）七铝酸十二钙
dolomite refractory brick 白云石耐火砖
dolomite refractory；dolomitic refractory 白云石耐火材料
dry vibratable refractory 干式耐火振捣料

E

electric furnace carbon brick 电炉碳砖
energy-saving coatings 节能涂料
environmentally friendly lightweight clay bricks 环保轻质黏土砖
expanded graphite；exfoliated graphite；EG 膨胀石墨
expanded perlite 膨胀珍珠岩
expanded vermiculite 膨胀蛭石

F

far-infrared coating 远红外涂料
fibrous porous zirconia ceramics 纤维多孔氧化锆陶瓷
fire clay 耐火黏土
fireclay bricks 耐火黏土砖
fireclay checker bricks 黏土格子砖
fireclay insulating board 耐火黏土绝热板
fireclay sleeve 黏土袖砖
fired magnesia brick 烧成镁砖
flake graphite 鳞片石墨
flame gunning refractory 火焰喷补耐火材料
floating bead 漂珠
floating bead ceramics 漂珠陶瓷
fly ash 粉煤灰
fly ash floating beads 粉煤灰漂珠
fly ash suspension 粉煤灰悬浮液
foam silicon carbide 泡沫碳化硅
forsterite refractory brick 镁橄榄石耐火砖
fused cast chrome-corundum bricks 熔铸铬刚玉砖
fused cast corundum bricks 熔铸刚玉砖
fused cast corundum refractory products 熔铸刚玉制品
fused cast zirconia refractory 熔铸氧化锆耐火

材料

fused magnesia 电熔镁砂（电熔氧化镁）

fused magnesia chrome brick; fused grain magnesia chrome brick 熔融镁铬砖

fused mullite 熔融莫来石

fused mullite particles 熔融莫来石颗粒

fused mullite powder 熔融莫来石粉末

fused zirconia-alumina-silica material (AZS) refractory 熔融氧化锆-氧化铝-二氧化硅耐火材料

fused cast zirconia-corundum bricks 熔铸锆刚玉砖

fused-cast AZS refractories 熔铸铝硅锆（AZS）耐火材料

fused-cast baddeleyite-corundum refractories 熔铸辉石刚玉耐火材料

G

glass wool 玻璃棉

GMH bricks for cement kilns 水泥窑用硅莫红砖

grade Ⅰ high alumina bricks Ⅰ等高铝砖

grade Ⅱ high alumina bricks Ⅱ等高铝砖

grade Ⅲ high alumina bricks Ⅲ等高铝砖

graphite annulus 石墨环

graphite anode; graphite cathode 石墨阳极

graphite articles 石墨（质）制品

graphite bar 石墨棒

graphite bath 石墨质熔池

graphite boat 石墨舟

graphite compression mould 石墨压模

graphite dust; graphite powder; powdered graphite; graphite crucible 石墨坩埚

graphite electrode 石墨电极

graphite fibers 石墨纤维

graphite fluoride 氟化石墨

graphite heater 石墨发热体

graphite marble 石墨大理石

graphite mortar 石墨火泥

graphite mould 石墨模型

graphite packing 石墨填料

graphite powder 石墨粉

graphite refractory 石墨（质）耐火材料

graphite sheet 石墨板

graphite tube 石墨管

graphite tuyere block 石墨风口砖

graphitic carbon nitride; C_3N_4 石墨相氮化碳

guimo bricks 硅莫砖

gunned refractory 喷补用耐火材料

gunning refractory 喷射耐火材料

H

heat radiative coating 热辐射涂料

high alumina bricks 高铝砖

high alumina bricks for casting ladle 盛钢桶用高铝砖

high alumina products 高铝质制品

high chrome oxide refractories 高氧化铬耐火材料

high density silica bricks 高密度硅砖

high emissivity coatings 高发射率涂层

high grog fireclay bricks 多熟料耐火黏土砖

high temperature infrared radiation energy saving coating 高温红外辐射节能涂层

high zirconia brick 高氧化锆砖

high zirconia fused-cast refractories 高氧化锆熔铸耐火材料

high-alumina refractory 高铝质耐火材料

high-alumina refractory bricks 高铝耐火砖

high-alumina stopper 高铝塞头砖

high-duty fireclay bricks 优质耐火黏土砖

high-refractory thermal insulating materials 高隔热耐火材料

high-temperature mould pressing carbon brick 高温模压碳砖

hollow spheres of refractory 耐火氧化物空心球

hot-blast stove silica bricks 热风炉硅砖

hydratable alumina bonded castable 可水合氧化铝结合浇注料

hypocrystalline zircon nanofibrous aerogels 亚晶锆石纳米气凝胶

I

injection mix 压注料
injection refractory 耐火压注料
insulation boards for steel casting 浇钢用隔热板
iron silicide nanoparticles 铁硅化物纳米颗粒

L

light forsterite brick 轻质镁橄榄石砖
lightweight refractory castable 轻质（隔热）耐火浇注料
lightweight Al_2O_3-MgO-C refractories 轻质 Al_2O_3-MgO-C 耐火材料
lightweight anorthite insulation bricks 钙长石轻质隔热砖
lightweight bricks 轻质砖
lightweight clay bricks 轻质黏土砖
lightweight high alumina bricks 轻质高铝砖
lightweight insulating fireclay bricks 轻质隔热黏土砖
lightweight insulation refractories 轻质隔热耐火材料
lightweight insulation silicon bricks 轻质隔热硅砖
lightweight magnesia 轻质氧化镁
lightweight microporous magnesia-based refractories 轻质微孔镁基耐火材料
lightweight mullite bricks 莫来石系轻质砖
lightweight mullite-based mullite-SiC bricks 轻质莫来石基莫来石-碳化硅砖
lightweight silica bricks 轻质硅砖
lime-bonded silica refractory 石灰结合硅质耐火材料
long nozzles 长水口
low carbon magnesia carbon brick 低碳镁碳砖
low carbon refractory 低碳耐火材料
low creep high alumina bricks 低蠕变高铝砖
low porosity fireclay bricks 低气孔耐火黏土砖

M

magnesia alumina chrome composite spinel bricks 镁铝铬复合尖晶石砖
magnesia brick 镁砖
magnesia calcia carbon brick 镁钙碳砖
magnesia carbon brick; MgO-C brick 镁碳砖
magnesia chrome brick; magnesia-chromite brick 镁铬砖
magnesia crucible 氧化镁坩埚
magnesia dry vibratable refractory 镁质干式振捣料
magnesia forsterite bricks 镁橄榄石砖
magnesia hercynite bricks 镁铁铝尖晶石砖
magnesia products 氧化镁制品
magnesia refractory products 氧化镁耐火材料制品
magnesia zirconia brick; magnesia-zirconia brick 镁锆砖
magnesia zirconia composite 氧化镁氧化锆复合材料
magnesia-alumina-carbon (MAC) bricks 镁铝碳砖
magnesia-alumina-titania bricks 镁铝钛砖
magnesia-based refractory 镁基耐火材料
magnesia-chrome refractory 镁铬耐火材料
magnesia-magnesium aluminate spinel refractory 氧化镁-镁铝尖晶石耐火材料
magnesia-silica brick; high-silica magnesite brick 镁硅砖
magnesia-spinel bricks 镁铝尖晶石砖
magnesia-spinel-zirconia brick 尖晶石镁锆砖
magnesia-zirconia co-clinker 氧化镁-氧化锆共混物
magnesia-zirconia refractory 镁锆耐火材料
magnesite-dolomite refractory; magnesia-dolomite refractory 镁质白云石耐火材料
metallurgical magnesia 冶金镁砂

metering nozzles 定径水口
MgO-ZrO_2 ceramic 氧化镁-氧化锆陶瓷
microcrystalline graphite 微晶石墨
micropore carbon brick 微孔碳砖
micropore corundum bricks 微孔刚玉砖
microporous calcium silicate 微孔硅酸钙
microporous lightweight insulator materials 微孔轻质隔热材料
microporous refractories 微孔耐火材料
microwave sintered zirconia 微波烧结氧化锆
molybdenum disilicide intermetalic 二硅化钼金属间化合物
molybdenum disilicide; $MoSi_2$ 二硅化钼
monolithic stopper 整体塞棒
mullite blocks 莫来石铸块
mullite bonded silicon carbide products 莫来石结合碳化硅制品
mullite bricks 莫来石砖
mullite ceramics 莫来石陶瓷
mullite chamotte 莫来石熟料
mullite-cordierite decks 莫来石-堇青石棚板
mullite-cordierite kiln furniture 莫来石-堇青石窑具
mullite-cordierite saggar 莫来石-堇青石质匣钵
mullite-corundum checker bricks 莫来石刚玉格子砖
mullite-corundum products 莫来石刚玉制品
mullite-corundum refractory 莫来石刚玉质耐火材料
mullite-SiC-O′-Sialon composites 莫来石-SiC-O′-Sialon 复合材料
mullite-zirconia refractories 莫来石-氧化锆质耐火材料

N

nano forsterite/nano magnesium aluminate spinel powders 纳米镁橄榄石/纳米镁铝尖晶石粉
natural diatomite 天然硅藻土
nitride bonded SiC 氮化物结合 SiC 制品

O

oxide bonded SiC 氧化物结合 SiC 制品

P

paraffin/expanded perlite material 石蜡/膨胀珍珠岩材料
periclase-forsterite lightweight heat-insulating refractories 方镁石-镁橄榄石轻质隔热耐火材料
periclase-magnesia alumina spinel bricks 方镁石-镁铝尖晶石砖
periclase-magnesium aluminate spinel refractory (PMAS) 方镁石-镁铝尖晶石耐火砖
phosphate-bonded castable 磷酸盐结合浇注料
phosphate-bonded castable 磷酸盐结合可塑料
phosphoric acid immersed fireclay bricks 磷酸浸渍耐火黏土砖
plastic refractory 耐火可塑料
polycrystalline alumina fiber 多晶氧化铝纤维
polycrystalline alumina-based fibers 多晶氧化铝基纤维
polycrystalline silicon carbide 多晶碳化硅
porous alumina ceramic 多孔氧化铝陶瓷
porous boron nitride 多孔氮化硼
porous calcium silicate 多孔硅酸钙
porous graphite 多孔石墨
porous lightweight clay bricks 多孔轻质黏土砖
porous plugs 透气元件
porous refractory thermal insulating materials 多孔隔热耐火材料
porous silicon carbide ceramics 多孔碳化硅陶瓷
prebaked carbon block 预焙碳块
presynthesized magnesia-zirconia 预合成氧化镁-氧化锆

R

ramming refractory 捣打料
reaction-bonded porous silicon nitride ceramics 反应烧结多孔氮化硅陶瓷

reaction-bonded silicon nitride 反应结合氮化硅

reaction-sintered zirconia-mullite composites 反应烧结锆莫来石复合材料

reactive sintered boron nitride 反应烧结氮化硼

rebonded electrically fused corundum bricks 再结合电熔刚玉砖

rebonded magnesia chrome brick 再结合镁铬砖

rebonded magnesia zirconia brick 再结合镁锆砖

rebonded magnesite brick 再结合镁砖

rebonded sintered corundum bricks 再结合烧结刚玉砖

reconstituted fused-grain magnesia-chrome refractory brick 再生电熔粒状镁铬耐火砖

refractory coating 耐火涂料（涂抹料）

refractory mortar 耐火泥浆

refractory mortar of heat setting 热硬性耐火泥浆

refractory mortar of hydraulic setting 水硬性耐火泥浆

refractory mud 耐火挤压料

refractory thermal insulation materials; insulation refractories 隔热耐火材料

resystallized silicon carbide products 重结晶 SiC 制品

roasted carbon brick 焙烧碳砖

rock wool 岩棉

S

self-baking carbon brick 自焙碳砖

self-bonded silicon carbide products 自结合 SiC 制品

semi graphite carbon brick 半石墨碳砖

semi-rebonded magnesia chrome brick 半再结合镁铬砖

semi-silica brick for coke oven 焦炉用半硅砖

semi-silica raw materials 半硅质原料

semi-silica refractories 半硅质耐火材料

semi-silicon carbide products 半 SiC 质制品

semi-stablized dolomite refractory brick 半稳定白云石耐火砖

Sialon bonded silicon carbide products 塞隆结合碳化硅制品

silica brick 硅砖

silica dry vibratable refractory 硅质干式振捣料

siliceous fly ash 硅质粉煤灰

silicon carbide ceramic matrix composite material 碳化硅陶瓷基复合材料

silicon carbide fiber 碳化硅纤维

silicon carbide film 碳化硅膜

silicon carbide nanosheets 碳化硅纳米片

silicon carbide nanotubes 碳化硅纳米管

silicon carbide nanowires 碳化硅纳米线

silicon carbide whiskers 碳化硅晶须

silicon dioxide bonded silicon carbide products 二氧化硅结合碳化硅制品

silicon infiltration reaction silicon carbide products 渗硅反应 SiC 质制品

silicon nitride bonded silicon carbide products 氮化硅结合碳化硅制品

silicon nitride ceramics; Si_3N_4 ceramics 氮化硅陶瓷

silicon nitride-based nanocomposites 氮化硅基纳米复合材料

silicon oxy nitride and compound phase nitride bonded silicon carbide products 氧氮化硅和复相氮化物结合碳化硅制品

silicon-boron-zirconium boride 硼化锆-硼化物

sillimanite bricks 硅线石砖

single crystal silicon carbide 单晶碳化硅

sintered (dead burned) magnesia 烧结镁砂（死烧镁砂）

sintered mullite products 烧结莫来石制品

sintered silicon nitride 烧结氮化硅

sintered zirconia ceramics 烧结氧化锆陶瓷

sintered zirconia mullite refractory products 烧结锆莫来石耐火制品

sintered zirconia refractory products 烧结氧化锆耐火制品

sliding nozzle; sliding gate 滑动水口

spalling resistant high-alumina bricks 抗剥落高铝砖
special refractory mortar 特种耐火泥浆
spinel-forsterite 尖晶石-镁橄榄石
stabilized dolomite refractory brick 稳定性白云石耐火砖
submerged entry nozzles 浸入式水口

T

tabular corundum 板状刚玉
taphole mud 有水炮泥
tar-bonded magnesite bricks 焦油结合镁砖
ternary boride-based cermet 三元硼化物基陶瓷
the diatomite composite hydrogel 硅藻土复合水凝胶
the submerged entry nozzles with swirling flow 旋涡浸入式水口
$TiZrO_4$ hollow spheres $TiZrO_4$ 空心球
transparent alumina ceramics 透明氧化铝陶瓷
tungsten silicide films 硅化钨薄膜

U

ultrafine silica powders 超细二氧化硅微粉
ultra-low cement refractory castable 超低水泥耐火浇注料
unfired high alumina bricks 不烧高铝砖
unfired high alumina bricks 免烧高铝砖
unfired lightweight clay bricks 不烧轻质黏土砖
unroasted thermal insulating refractory materials 不烧隔热耐火材料

W

waterless mud 无水炮泥
wearing resistant refractory castable 耐磨耐火浇注料

Z

zircon corundum brick 锆刚玉砖
zircon ramming refractory 锆英石质耐火捣打料
zirconia fiber 氧化锆纤维
zirconia products 氧化锆制品
zirconia refractory 氧化锆耐火材料
zirconium boride coating 硼化锆涂层
zirconium boride particles 硼化锆颗粒
zirconium boride-nichrome 硼化锆-铬镍合金
zirconium mullite 烧结锆莫来石砖

*

β-Sialon bonded corundum bricks β-Sialon 结合刚玉砖
β-Sialon bonded corundum-SiC composites β-Sialon 结合刚玉碳化硅复合材料
β-SiC bonded silicon carbide products β-SiC 结合 SiC 制品